Springer Theses

Recognizing Outstanding Ph.D. Research

Aims and Scope

The series "Springer Theses" brings together a selection of the very best Ph.D. theses from around the world and across the physical sciences. Nominated and endorsed by two recognized specialists, each published volume has been selected for its scientific excellence and the high impact of its contents for the pertinent field of research. For greater accessibility to non-specialists, the published versions include an extended introduction, as well as a foreword by the student's supervisor explaining the special relevance of the work for the field. As a whole, the series will provide a valuable resource both for newcomers to the research fields described, and for other scientists seeking detailed background information on special questions. Finally, it provides an accredited documentation of the valuable contributions made by today's younger generation of scientists.

Theses may be nominated for publication in this series by heads of department at internationally leading universities or institutes and should fulfill all of the following criteria

- They must be written in good English.
- The topic should fall within the confines of Chemistry, Physics, Earth Sciences, Engineering and related interdisciplinary fields such as Materials, Nanoscience, Chemical Engineering, Complex Systems and Biophysics.
- The work reported in the thesis must represent a significant scientific advance.
- If the thesis includes previously published material, permission to reproduce this must be gained from the respective copyright holder (a maximum 30% of the thesis should be a verbatim reproduction from the author's previous publications).
- They must have been examined and passed during the 12 months prior to nomination.
- Each thesis should include a foreword by the supervisor outlining the significance of its content.
- The theses should have a clearly defined structure including an introduction accessible to new PhD students and scientists not expert in the relevant field.

Indexed by zbMATH.

Maxime Debertolis

Quantum Impurity Problems in the Framework of Natural Orbitals

A Comprehensive Study

Doctoral Thesis accepted by
The Université Grenoble Alpes, Grenoble, France

Springer

Author
Dr. Maxime Debertolis
Physics Institute
University of Bonn
Bonn, Germany

Supervisor
Dr. Serge Florens
Institut Néel
The Université Grenoble Alpes
Grenoble, France

ISSN 2190-5053 ISSN 2190-5061 (electronic)
Springer Theses
ISBN 978-3-031-47232-9 ISBN 978-3-031-47233-6 (eBook)
https://doi.org/10.1007/978-3-031-47233-6

This Springer imprint is published by the registered company Springer Nature Switzerland AG
The registered company address is: Gewerbestrasse 11, 6330 Cham, Switzerland

Paper in this product is recyclable.

Supervisor's Foreword

It is a real pleasure to provide an introduction to this book presenting the Ph.D. thesis of Maxime Debertolis. Maxime joined the Néel Institute at CNRS-Grenoble in 2019, strengthening an established collaboration with Prof. Izak Snyman at the University of Witwatersrand.

The general framework of Maxime's Ph.D. topic concerned the simulation of quantum systems made of many particles, a challenging physical problem with many applications for the understanding of correlated materials and quantum devices. Maxime demonstrated that a huge compression of quantum states could be achieved for a peculiar class of electronic models called quantum impurities, using the concept of natural orbitals borrowed from quantum chemistry. As a result, the ground state of the famous single impurity Anderson Hamiltonian, previously thought to be many-body in nature, turned out to be accurately described by a few-body Ansatz on the basis of natural orbitals.

This research opens a clear path for novel advances in the field, building on an iterative algorithm specially designed by Maxime, that efficiently generates natural orbitals for generic classes of impurity models. In his thesis, Maxime was able to address for the first time the complex spatial charge and spin correlations around a quantum impurity in the presence of disorder and on large-scale lattices. The perspectives of Maxime's work are numerous, for instance by harnessing time-dependent natural orbitals in order to tackle the dynamics of open quantum systems, a notoriously difficult problem limited by the entanglement barrier.

The manuscript of Maxime's Ph.D., now published as a Springer thesis, contains also numerous details on the numerical implementation and careful testing of the methodology, that will hopefully help the readers to explore these new concepts for their own research.

Grenoble, France
May 2023

Dr. Serge Florens

Abstract

Strong correlations arising from interactions between particles are responsible for many phenomena in condensed matter physics, from superconductivity to heavy fermions. An accurate theoretical description of such complex systems requires to solve so-called many-body problems, which are inherently difficult to tackle due to the exponential growth of the underlying Fock space with the size of the system. Usually, single-particle descriptions or perturbation theory at first order fail to provide satisfactory results in correlated regimes, and pertinent Ansätze or cutting-edge numerical methods are required to solve these problems.

Among the wide variety of strongly correlated electronic systems, quantum impurity systems are of central interest. These systems, where a few localized degrees of freedom that undergo strong interactions hybridize with a larger system of free particles, have been widely studied for decades. Beyond their relevance to study dilute dynamic impurities in metals, quantum dots, or more complex nanostructures, they are exploited in a broader class of problems through dynamical mean field theory (DMFT), that is designed for fully interacting lattice problems. Despite the effort devoted to studying of these problems, some questions remain open and new approaches are required to address them.

In this thesis, quantum impurity problems are tackled with the use of natural orbitals, a representation rooted in the framework of quantum chemistry. The numerical renormalization group, the standard non-perturbative method to treat these systems, is employed to demonstrate the hierarchical structure of correlations unveiled by natural orbitals. An ansatz for the wave function of the ground state is then proposed, separating explicitly a small set of correlated orbitals to a large uncorrelated one, leaving the problem to a few-body complexity. This simplification brought new insights for simulating quantum impurity problems, and a new algorithm is developed to generate an optimized set of correlated orbitals independently of existing methods, going beyond their usual limitations.

In the first place, the algorithm is used to compute the screening cloud in large one-dimensional chains and in a realistic 2D square lattice coupled to a localized adatom. This observable is of great interest to understand the screening mechanism of the magnetic moment of the impurity that plays a central role in transport properties, for unidimensional systems or in bulk metals experiencing RKKY interactions. Spatial correlations were found to be highly non-symmetric, breaking the s-wave picture. In one direction, correlations vanish way faster than the usual isotropic correlation length, while, in the other one, they spread further than this same expected length. This anisotropy suggests that impurities in bulk systems may interact differently depending on their relative positions with respect to the lattice.

Taking further advantage of the efficiency of this new algorithm allowed us to perform a statistical study on large systems to investigate correlations around a quantum impurity in a disordered host, averaging over a large amount of realizations, which was not possible with previous approaches. The question motivating this study was the fate of the correlation cloud spreading into the disordered electronic bath. Although spatial correlations are expected to vanish when the disorder increases, long-range correlations survive to some configurations of the random potential, inducing a bimodal structure of probability distribution functions of relevant scales. We also observe different effects on correlations depending on whether the disorder is applied in the charge or spin channel.

Finally, these disordered problems are investigated in the non-interacting situation with random matrix theory, which reproduced qualitatively the results obtained before. The joint distribution of one-particle orbitals energies and amplitudes is calculated, which allows to compute in principle any disordered averaged local correlation functions, and calculations in the large-N limit and using a random two by two matrix toy model are presented, which reproduces all qualitative aspects of disordered quantum impurity models.

Publications related to this thesis

1. Debertolis, M., Florens, S. & Snyman, I. Few-body nature of Kondo correlated ground states. *Phys. Rev. B* **103**, 235166 (23 June 2021).
2. Debertolis, M., Snyman, I. & Florens, S. Simulating realistic screening clouds areound quantum impurities: Role of spatial anisotrophy and disorder. *Phys. Rev. B* **106**, 125115 (12 Sept. 2022).

Acknowledgements

The writing of this thesis manuscript was the culmination of three years of work at the Néel Institute, encumbered by almost two very special years of successive confinements due to the epidemy of COVID-19. Tackling the most advanced problems that I ever had to solve in such conditions was an additional challenge, and I could not have done it alone. It is therefore with great pleasure that the first pages of this manuscript go to all those who, in their own way, have participated in the accomplishment of this work.

First of all, I thank the rapporteurs Denis Ullmo and Sebastien Burdin for having carried out this task and attended the defense. I thank by the same occasion the examiners, Hermann Sellier, Denis Feinberg, Sabine Andergassen and Rok Žitko for their interest in my work.

As the person who had the most impact on my scientific learning, I am very grateful to Serge for having directed my work during these three years. You were always present to answer my questions, whether they concerned some technical details or the deeper understanding of physical phenomena. I thank you for the freedom you gave me in the work we did, and for letting me get interested and involved in projects that motivated me.

Although he was not present on a daily basis as Serge was, I would like to thank Izak for all the discussions we had during most of my projects. You were always able to bring a new eye to unlock new and relevant ways of thinking. Thanks to you, from a more technical point of view, I was able to develop my skills, on the numerical level, beyond some naive implementations that I developed alone.

Thank you to all the people in the M-building who were there when we were able to set foot in the lab again for being there every day. For all the various and scientific discussions, for all the hours (that I dare not even count) spent drinking coffee or playing babyfoot. For all this, thank you to Quentin, Sergueï, Théo, Guillaume, Baptiste, Justin, Hadi, Nant, Selma and Pierrick.

Des remerciements spéciaux vont tout d'abord à Théo, *le grand frère*, d'avoir été là pendant ces deux premières années, au laboratoire comme aux différentes écoles auxquelles nous avons assistées ensemble. Ce fut toujours un plaisir de discuter avec toi, de profiter de tes connaissances en physique qui m'ont toujours impréssionnées, et de pouvoir me plaindre de quoi que ce soit à la moindre occasion. Merci aussi particulièrement à Guillaume d'avoir été un co-bureau génial quand tu étais là . Ma productivité te remercie moins, mais je n'aurais pas pensé passer de si bon moments au laboratoire, ni que j'allais un jour chanter le générique d'intervilles avec un accent québecois en calculant des fonctions de Green. N'oublie pas de bosser les échecs et le baby, c'est quand même la dedans que l'on est meilleurs.

En dehors du laboratoire, merci à tous ceux qui ont été là depuis maintenant bien plus longtemps que le début de cette thèse. Une liste exhaustive serait dure à mettre en place, vous avez tous été un énorme soutien depuis tout ce temps, et avez toujours contribué à me sortir la tête de tous les problèmes que je pouvais rencontrer au quotidien.

Je remercie tout particulièrement mes parents, mon frère, et toute ma famille de m'avoir soutenu depuis le début, de m'avoir compris dans ma passion pour le monde de la recherche en physique, et de m'avoir accompagné dans tout ce que j'ai pu entreprendre en dehors du milieu scolaire. Un merci spécial va à mes parents pour avoir complètement géré le pot de thèse, qui fut très réussi et que, étant donnée ma gestion des timings, je n'aurais jamais pu organiser seul. Merci aussi à ma belle famille pour l'accueil formidable qu'ils m'ont toujours apporté, et d'avoir assisté à ma soutenance.

Enfin, merci à Manon, pour toutes ces années, pour tout ce que tu as pu supporter, d'avoir toujours été à mes côtés. Merci de m'avoir aidé à garder les pieds sur terre et à prendre du recul sur mon travail et mes façons de penser, je suis persuadé que tu m'as rendu meilleur, et que je ne serais pas là où j'en suis sans toi.

Contents

Part I Introduction

Part II Quantum Impurity Problems and Natural Orbitals

Part I
Introduction

Chapter 1
The Quantum Impurity Problem

The frontier of science extends all along a long line from the newest and most modern intensive research, over the extensive research recently spawned by the intensive research of yesterday, to the broad and well developed web of extensive research activities based on intensive research of past decades.

P. W. Anderson, Science (1972)

1.1 The Many-Body Problem in Condensed Matter

Giving a clear and appropriate description of matter requires to consider the coexistence of an astronomically large number of particles. A piece of metal is expected to host no less than 10^{23} particles, and trying to follow the behavior of so many objects interacting simultaneously has more to do with madness than physics. At some point I have been introduced to many-body quantum physics, and how powerful mathematical tools combined to *simple* physical models can explain phenomena originating from the interaction of all these particles. Employing the word simple here is a bit crude, since building such theory requires a deep and accurate physical understanding of what may happens in these quantum systems so far from our sensible world. In the end, what matters is the smallest number of ingredients that are required for a given phenomena to emerge, and how is it affected by this or that disturbance. Then, when we believe that the model that we built could explain the physics in our system (or any new phenomenon that might exists under some specific conditions), we have to arrange the problem such that we can compute any *observable* (a quantity that can be measured) of interest. The models are defined through a specific Hamiltonian operator, and can be solved in principle by finding the eigenstates of the latter: in that

M. Debertolis, *Quantum Impurity Problems in the Framework of Natural Orbitals*, Springer Theses, https://doi.org/10.1007/978-3-031-47233-6_1

sense, the Schrödinger equation is solved, and we have ideally access to the desired physical quantity of the system. There exists plenty of good references covering the many-body problem in quantum mechanics and its applications, but I suggest here the ones that I mostly followed during my PhD [1–4].

When the Hamiltonian is quadratic in the fields of the particles under study, it can be diagonalized in the Hilbert's space that is spanned by a set of N orbitals $|\phi_n\rangle = c_n^\dagger|0\rangle$ in which the problem is defined. We will abuse of the word *orbital*, that will in the following denote an accessible state for each particle, in a wider sense than atomic or molecular orbitals commonly employed in quantum chemistry. We introduced the creation operator $c_n^\dagger$, defined in the second quantization representation of the physical fields, which creates a particle in the one-particle state labelled by n. In this thesis, we will focus on fermionic particles, whose principal properties are the fermionic anticommutation relation $\{c_n^\dagger, c_m\} = \delta_{mn}$ and the Pauli principle $(c_n^\dagger)^2|0\rangle = 0$, which forbids two fermions to be in the same quantum state. Note that these properties are general and do not depend on the Hamiltonian. This one-particle modelisation can be used to describe free particles and can explain localization effects in disordered systems, the integer quantum hall effect, etc. It brings accurate results when the effect of inter-particle interactions is negligible, which is a strong statement in condensed matter.

Most of the time, interaction between particles must be taken into account, and when introducing the typical density-density Coulomb interaction between charged particles, the Hamiltonian is no more quadratic in the fields and the problem is usually treated in Fock space. This space is defined as the direct sum of single particle Hilbert spaces for any number of particles. Formally, the Fock space is defined as: $\mathcal{F} = \bigoplus_{m=0}^{N} \mathcal{H}^{\otimes m}$. Usually, the occupation number representation for a state is used in this space, so that each different state corresponds to a specific (unique) distribution of n particles within the N accessible orbitals. A Fock state can thus be written $|\Phi_a\rangle = |n_1, \ldots, n_i, \ldots, n_N\rangle = (c_1^\dagger)^{n_1} \ldots (c_i^\dagger)^{n_i} \ldots (c_N^\dagger)^{n_N}|0\rangle$, with $\sum_{i=0}^{N} n_i = m$, where n_i denotes the occupation of orbital i and m the number of particles in the state. Thus, the Fock space is spanned by all Fock states built by all possible configurations of $n_{i=0,\ldots,n} \in \{0, 1\}$, and is of dimension $\mathcal{D} = 2^N$. The exponential growth of the Fock space with the number of accessible orbitals makes the study of such systems difficult, and clever approximations must be done to reduce the complexity of the problem, for instance by focusing on a subspace that gives a good description of the whole problem.

As soon as interactions play a central role in the physics at play, the system is said to be strongly correlated: correlations between all degrees of freedom reveal emergent phenomena, that could not be explained by a simple extension of a single-particle solution of the free problem. Among the most notable effects coming from strong fermionic correlations are superconductivity, Mott insulators, Fermi liquids, etc.

These problems can be tackled by many-body analytical methods including Green's functions, perturbation theory, Hartree-Fock theory, [1–3, 5] etc., or numerical simulations, from quantum Monte-Carlo [6, 7] to renormalization group algorithms [8, 9]. Still, a lot of questions are left open in systems where every existing tool fails to provide satisfactory results, making the theoretical study of strongly correlated systems one of the most active field of condensed matter.

We will be focused on the rest of this part on quantum impurity models, describing a few degrees of freedom coupled to a large bath of non-interacting particle via Coulomb interaction, while hybridizing with the latter. These seemingly simple many-body problems exhibit non-trivial phenomena emerging from strong correlations, and are widely studied nowadays for the description of qubits or more complex lattice problems. We will start with a brief review of the Kondo problem, through which quantum impurity models were introduced, and then introduce the models that will be study in the following of the manuscript.

1.2 The Kondo Effect in a Nutshell

The following story is well known today, and hence we will not dwell on too specific details here. For a thorough review, the reader can refer to Refs. [2, 5, 10]. The premises of the large interest in quantum impurity problems lies in an anomaly observed in the resistivity of some metals at low temperatures. The processes responsible of the temperature dependence of the resistivity were well described at this time by solid state physics [11]. As stated by Matthiessen's rule, the scattering of conduction electrons by phonons—collective displacements of atoms in the metal—were supposed to disappear gradually when lowering the temperature, with the progressive freezing of phonons. As a result, the resistivity was expected to vanish as a power law of the temperature, up to some constant depending on the concentration of static defects in the lattice.

What was actually observed by W.J. de Haas and his collaborators [12] went against these predictions: in Au wires, below a certain temperature, the resistivity was starting to rise again. It has been understood later that magnetic impurities were responsible of this effect, but the mechanism was not clear. P.W. Anderson, following ideas of J. Friedel concerning virtual bound states [13] between electrons and magnetic scatterers, built the first minimalist theoretical model describing this effect, which we know today as the Anderson impurity model [14]. However, J. Kondo introduced a simpler s-d model [15], or Kondo model, involving a localized spin $\mathbf{S}$ coupled via exchange interaction J to conducting states. It appeared that this model is only a low-energy description [16] of the more general Anderson impurity model, that will be presented with more details in the following. Then, Kondo performed his now notorious perturbation theory that explained the logarithmic increase of the resistivity with temperature.

However the perturbative development was diverging as the temperature was approaching zero, so that the solution was not satisfactory. To address this limit, several theoretical studies appeared in a short period of time: Abrikosov-Migdal [18] and Fowler-Zawadowski [19] used the renormalization group at some perturbative order, and Anderson introduced his *Poor man's scaling* [17] to understand the low

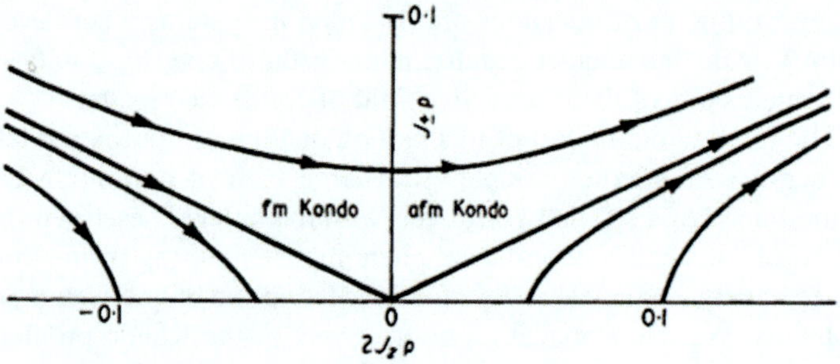

Fig. 1.1 Set of hyperbolic lines connecting equivalent problems [17] for the anisotropic Kondo model. The straight lines account for the SU(2) symmetric Kondo model with only one coupling channel J. The ferromagnetic case is asymptotically equivalent to the $J_\pm = 0$ problem, corresponding to the solvable Ising spin in a free electronic gas. In other cases, one must consider the strong coupling limit $J \to \infty$

energy physics in terms of equivalent problems. It turns out that as the temperature (or the considered energy scale) is lowered, interactions between the localized spin and free electrons strengthen, such that below an energy scale T_K, the effective coupling $\tilde{J}$ gets so strong that a possible picture is that of a singlet state, in such a way that the Kondo impurity is reduced to a non-magnetic scatterer, see Fig. 1.1. These arguments elucidated the failure of perturbation theory as $T \to 0$ ($\tilde{J} \to \infty$), which is therefore only valid for high $T \gg T_K$.

A few years later, K.G. Wilson came with a powerful non-perturbative algorithm that closed the debate [9]. The numerical renormalization group (NRG), successively transforming the Hamiltonian parameters until they reach a fixed point, was a breakthrough for the field of Kondo physics, and in general for numerical simulations of many-body Hamiltonians. It is still a commonly used method to simulate quantum impurity models, and a detailed description of the algorithm will be provided later on. The real power of the method is its ability to describe the full crossover between high energy properties and the strong coupling limit at low energy. Wilson provided a definitive answer for the ground state of the spin-$\frac{1}{2}$ problem, and showed that the ratio χ/γ of the susceptibility χ of the impurity and of the specific heat coefficient γ was twice that of free electrons. It suggests that the behavior is that of free particles, with some renormalized parameters. This opened the door to P. Nozières, that used the results of the strong coupling limit to show that this fixed point behaves as a *local* Fermi liquid [20], the screening of the impurity spins yielding to a remanent interaction between electrons in the vicinity of the spin.

The main line of this story ended with the analytical solution of the Kondo model for $S = \frac{1}{2}$. Andrei-Lowenstein [21] and Wiegmann [22] used the Bethe ansatz to find the spectrum and wave functions of the Kondo and Anderson models. Analytical results for the thermodynamics at all temperatures were found, and were in agreement with Wilson's results. The results are also valid in higher than one dimension, if the scattering can be assumed to be rotationally invariant (s-wave scattering). The models are integrable under one main assumption: the dispersion relation of electrons must be linearized at all scales In this sense, the results tend to describe the universal regime $T \ll D$, but any effect related to the band structure is not described [23].

As a small disgression, I would like to discuss some implications of these results in terms of thermodynamics. An integrable system is defined in analytical mechanics through an infinite number (scaling with the number of degrees of freedom at least) of independent constant of motion, such that the system is under strong constraints, and usually does not relax to thermal equilibrium, i.e. does not thermalize. However, any realistic description of Kondo models requires a finite system, with a separation of energy scales that is not as perfect as assumed by the Bethe ansatz solution. It has been actually shown with NRG simulations that these model actually satisfy the eigenstate thermalization hypothesis (ETH) [24], even if these results must be understood with a lot of care, since the bath discretization that is required for these simulations may not be able to describe long time dynamics [25] required to explore thermalization properties.

1.3 Models of Quantum Impurities

Throughout the early history of Kondo impurities, two models were introduced to describe single magnetic impurities in electronic hosts: the Kondo and the Anderson models. However these models apply to wider applications than the one presented until now. They provide a good description of quantum dots interacting with one or more metallic leads. These nanostructures in which the parameters at play can be highly adaptable have started to be designed in the early 2000's. These structures allow to simulate atoms [26] in different regimes, also acting as electronic transistors: the onsite Coulomb repulsion energy must be paid if we want another electron to occupy the dot. Hence the number of electrons in the dot is controllable, switching from conducting to insulating states (Coulomb blockade) as the gate voltage is tuned, corresponding to the onsite energy of the impurity [27]. In the Coulomb blockade regime, an odd number of electrons can be in the dot, thus holding a magnetic moment. In this configuration, the screening process of the impurity can be examinated by tuning the microscopic parameters, which makes these objects testbeds for Kondo physics [28, 29].

In another context, quantum impurity problems provide the underlying physics of interacting lattice problems in a local perspective. Dynamical mean field theory [30] builds the local Green's function by solving a correlated Anderson impurity model, whose parameters are obtained self-consistently. This method greatly contributed to the field of strongly correlated materials, from physical models to realistic materials with quantum chemistry embedded techniques [31]. A true intellectual effort has been implemented to solve these impurity models in the most efficient way, allowing better quantitative agreements to more and more complex and realistic problems.

In this section, we will present the single impurity Anderson model (SIAM), and how it is related to other widely used models. We aim to introduce the interacting resonant level model (IRLM) that has been under study during this work, which shares the same low energy properties of the SIAM and Kondo models.

1.3.1 Single Impurity Anderson Model

As stated before, the SIAM is built to describe the interaction between an atom with a partially filled d or f orbital thus carrying a magnetic moment, and a metal represented by a bath of non-interacting conduction electrons. The Hamiltonian in second quantization reads:

$$\mathcal{H}_{\text{SIAM}} = \sum_{\sigma} \epsilon_d d_\sigma^\dagger d_\sigma + U n_\downarrow n_\uparrow + \sum_{k,\sigma} \epsilon_{k\sigma} c_{k\sigma}^\dagger c_k + \sum_{k,\sigma} (V_k\, c_{k\sigma}^\dagger d_\sigma + \text{h.c.}). \quad (1.1)$$

The operator $d_\sigma^\dagger$ (d_σ) creates (annihilates) an electron of spin $\sigma \in \{\uparrow, \downarrow\}$ in the localized orbital, while the operator $c_{k\sigma}^\dagger$ creates an electron in the conduction band with a wave vector k and spin σ. The Hamiltonian is divided in three parts, as it is common for impurity models: an impurity part, with U the local Coulomb interaction (taken postive to be repulsive between electrons) that is paid when two electrons with opposite spin occupy the impurity, and ϵ_d the chemical potential of the orbital. ϵ_d is defined with respect to the Fermi energy of the conduction band, that is taken to be $E_\text{F} = 0$. If $\epsilon_d < 0$, the impurity will be more likely occupied, and empty if $\epsilon_d > 0$. A bath part, defined by the dispersion relation of conduction electrons ϵ_k. A coupling part, represented by a tunnel coupling V_k between the impurity and a state in the conduction band of energy ϵ_k. The density-density term, describing the interaction, is quartic in the fermionic fields, which makes this problem truly many-body when the interaction is not negligible compared to the other parameters. Note that natural units are used trough the manuscript, setting e, $\hbar$ and k_B equals to 1, and unless otherwise stated the energies will be in units of D, the half-bandwidth of the bath density of states.

This model exhibits two simple but not trivial limits given by the strength of the interaction U: the non-interacting case $U \to 0$ and the so called *atomic* limit $U \to \infty$. We will discuss some properties of the non-interacting case in a later paragraph with the introduction of the IRLM. We focus first on the atomic limit, that is closely linked to the Kondo model.

1.3.2 Atomic Limit of the SIAM

When the interaction becomes the dominant parameter, the coupling and bath parts can be neglected. In this way, only the impurity part remains, and constitutes an isolated system. There are four states available in this limit: one without any electrons on the impurity $|0\rangle$, that has an energy $E_0 = 0$; two degenerate states with occupation $\sum_\sigma n_{d\sigma} = 1$ on the impurity $d_\downarrow^\dagger|0\rangle$ and $d_\uparrow^\dagger|0\rangle$ of energy $E_1 = \epsilon_d$; a last state with $\sum_\sigma n_{d\sigma} = 2$, written $d_\uparrow^\dagger d_\downarrow^\dagger|0\rangle$, with $E_2 = U + 2\epsilon_d$. At fixed ϵ_d and $U \gg |\epsilon_d|$ it is clear that $E_2 \gg E_1, E_0$, such that double occupancy on the dot is a highly non-probable process. It is also evident that $\epsilon_d < 0$ is required for the ground state to hold a magnetic moment. This simplified picture tells us about the separation of energy scales as a sketch of the true problem. These arguments are somehow equivalent to the decoupled system $V_k = 0$, and to explore beyond the atomic limit, V_k is considered as a small perturbation in front of U and ϵ_d.

1.3.3 Schrieffer-Wolff Transformation and the Kondo Model

The full correspondence between the Kondo-like behavior and the Anderson model has been given by Schrieffer and Wolff [16], by mapping the low energy sector of the SIAM to an effective model, that turned out to be that studied by Kondo for the resistance increase in metals. As explained before, the coupling term V_k will be treated as a perturbation, which is equivalent to a projection onto the low energy sector defined in the atomic limit. However, it is not possible to treat this problem at first order in V_k: the projection would be trivial, since the term V_k couples the two sectors, and will simply vanish. The perturbation theory does not work at higher orders, since some terms are diverging in the expansion in V. Hence, the idea behind the transformation is to perform a canonical transformation $T = e^S$ after which the sectors are properly separated. The problem to consider is:

$$\tilde{\mathcal{H}} = e^S\, \mathcal{H}_{\text{SIAM}}\, e^{-S}. \tag{1.2}$$

In this way, the first order terms in V_k are canceled explicitely and the sectors well separated. We expand Eq. (1.2) using the Baker-Campbell-Haussdorf formula, and we label by H_1 the term involving V_k in the Hamiltonian, and by H_0 all remaining terms:

$$\tilde{\mathcal{H}} = H_0 + H_1 + \left[S, H_0\right] + \frac{1}{2}\left[S, H_1\right] + \frac{1}{3}\left[S, \left[S, H_1\right]\right] + \dots \,. \tag{1.3}$$

This form is used when H_1 is off-diagonal in the basis of H_0 which can be assumed here, and the wanted transformation consists in cancelling the linear terms in V in $\tilde{\mathcal{H}}$. The equation to be solved is thus:

$$\left[S, H_0\right] = -H_1\,. \tag{1.4}$$

In order to find S, the problem is first simplified. In the non-interacting case, a guess for S_0 would be in a similar form to the term to be cancelled, namely H_1:

$$S_0 = \sum_{k,\sigma} A_{k\sigma} \left(c^\dagger_{k\sigma} d_\sigma - d^\dagger_\sigma c_{k\sigma} \right). \tag{1.5}$$

Here S_0 is guaranteed to be anti-hermitian, imposing T to be unitary as desired. Equation (1.4) is solved and once the commutator is computed, it is plain to identify $A_{k\sigma} = V_k / (\epsilon_k - \epsilon_d)$. In the interacting case, the interaction $U n_\downarrow n_\uparrow$ is added to H_0, and the commutation relation is not filled by S_0 anymore:

$$\left[S_0, H_0 \right] = -H_1 - U \sum_{k,\sigma} A_{k\sigma} (c^\dagger_{k\sigma} d_\sigma + d^\dagger_\sigma c_{k\sigma}) n_{d-\sigma}. \tag{1.6}$$

The additional term involving $n_{d-\sigma}$ has to be canceled, so that a new guess for S in the interacting case reads:

$$S = S_0 + \sum_{k,\sigma} B_{k\sigma} \left(c^\dagger_{k\sigma} d_\sigma + s'\, d^\dagger_\sigma c_{k\sigma} \right) n_{d-\sigma}. \tag{1.7}$$

Proceeding as before, $B_{k\sigma} = (U V_k)/((\epsilon_k - \epsilon_d - U)(\epsilon_k - \epsilon_d))$. Putting everything together, the expression given in the seminal article is found and reads:

$$S = \sum_{k,\sigma} \frac{V_k}{\epsilon_k - \epsilon_d} \left(c^\dagger_{k\sigma} d_\sigma - d^\dagger_\sigma c_{k\sigma} \right) \left(1 - \frac{U n_{d-\sigma}}{\epsilon_k - \epsilon_d - U} \right). \tag{1.8}$$

Although the expression for the generator of the transformation has been found, $\tilde{\mathcal{H}}$ is still an infinite series of terms. H_1 being treated as a perturbation in the case of interest, it is approximated by the first non-vanishing terms in V_k, reading:

$$\tilde{\mathcal{H}} \simeq H_0 + \frac{1}{2} \left[S, H_1 \right]. \tag{1.9}$$

This second term is computed directly with Eq. (1.7), and, for convenience, the following spinor notation is used:

$$\Psi_k = \begin{pmatrix} c_{k\uparrow} \\ c_{k\downarrow} \end{pmatrix}, \quad \Psi_d = \begin{pmatrix} d_\uparrow \\ d_\downarrow \end{pmatrix}. \tag{1.10}$$

The Hamiltonian is developed, and four terms are found: H_{ex}, H_{dir}, $H_0^{'}$ and H_{ch}, to keep the original notation. However, most of these terms are not relevant in the limit that is considered. Indeed, H_{ch} only changes the number of electrons on the impurity by two, such that it only lies in the high energy sector and disappears after projection. $H_0^{'}$ contains the same operators as H_0, such that it is only an energy shift for U and ϵ_d. H_{dir} allows no spin-flips, and since $\Psi^\dagger_d \Psi_d = 1$ in the low energy sector, it only acts as a local potential in the sea of conduction electrons and is absorbed in the definition of ϵ_k. Finally, only the H_{ex} part remains, which represents an exchange interaction between the localized moment and conduction electrons:

$$H_{ex} = \sum_{kk'} J_{k'k} \left(\Psi^\dagger_{k'} \frac{\boldsymbol{\tau}}{2} \Psi_k\right) \cdot \left(\Psi^\dagger_d \frac{\boldsymbol{\tau}}{2} \Psi_d\right), \tag{1.11}$$

with $J_{k'k} = -UV_k V_{k'} \left(\left[(\epsilon_k - \epsilon_d - U)(\epsilon_k - \epsilon_d)\right]^{-1} + \left[(\epsilon_{k'} - \epsilon_d - U)(\epsilon_{k'} - \epsilon_d)\right]^{-1}\right)$, and $\boldsymbol{\tau}$ are the Pauli matrices. In addition, the interaction term in H_0 also vanishes, such that the Kondo model is recovered:

$$\mathcal{H}_{\text{Kondo}} = \sum_{k,\sigma} \epsilon_k c^\dagger_k c_k + \frac{J}{2} \sum_{kk'\sigma\sigma'} \mathbf{S}_d \cdot \left(c^\dagger_{k\sigma} \boldsymbol{\tau} c_{k'\sigma'}\right). \tag{1.12}$$

In this model, the momentum dependancy of the exchange interaction can also be neglected because only the physics close to the Fermi energy is studied, taking $k \sim k' \sim k_F$, and hence $J = -2V^2_{k_F} U \left[\epsilon_d \left(\epsilon_d + U\right)\right]^{-1}$. The impurity is only coupled to the fermionic field at the origin, defined as $1/\sqrt{N} \sum^N_{k\sigma} c_{k\sigma} = c_0$. In the present limit where $\epsilon_d \ll -D$ and $\epsilon_d + U \gg D$, the exchange interaction is antiferromagnetic as predicted by Kondo calculations for metals. As stated before, there is an energy scale T_K under which perturbation theory does not apply, and it is defined from the above parameters as $T_K \propto D e^{-\frac{1}{\rho_0 J}}$, where ρ_0 is the density of states at k_F of the band states. The quench of the impurity spin by the conduction electrons involves states with energy $\epsilon_k \ll D$, such that the description is efficient when the effective exchange coupling is small and $T_K \to 0$, or equivalently $\rho_0 J \ll 1$. This corresponds to the strong coupling (or strongly correlated) regime that we discussed before, in which the relevant energy scale is exponentially small, and the physics can be equivalently described by the two models presented above.

1.3.4 Bosonization and the Interacting Resonant Level Model

Previously to the Bethe ansatz solution, some efforts were put in finding a mapping of the Kondo model to a solvable one [32–36]. It was first seen that certain quantities were equivalent between the Kondo model and effective non-interacting problems, for some specific values of the interaction J. At this time, the equivalence was cumbersome to establish, and the bosonization method introduced some years later provided a simple yet rigorous mapping between these models [35, 37]. We will give the general idea behind the bosonization transformation that interests us following Ref. [3], without entering in deep mathematical details that are out of the scope of this thesis. For an exhaustive discussion, see Ref. [38, 39].

The starting point is the spin anisotropic Kondo model (AKM):

$$\begin{aligned}\mathcal{H}_{\text{AKM}} &= \sum_{k,\sigma} \epsilon_k c^\dagger_{k\sigma} c_{k\sigma} + \frac{J_z}{2}\left(\psi^\dagger_\uparrow(0)\psi_\uparrow(0) - \psi^\dagger_\downarrow(0)\psi_\downarrow(0)\right) S^z_d \\ &+ \frac{J_{xy}}{2}\left(\psi^\dagger_\uparrow(0)\psi_\downarrow(0) S^-_d + \psi^\dagger_\downarrow(0)\psi_\uparrow(0) S^+_d\right).\end{aligned} \tag{1.13}$$

where S^z is the spin of the impurity, and $S^\pm_d = S^x_d \pm i S^y_d$ are the impurity spin ladder operators, basically performing spin flips. The local fermionic operators $\psi_\sigma(0) = N^{-1/2}\sum_k^N c_k$ are introduced, through which the impurity only explicitely couples to the first site of the 1D chain representing the bath. The AKM is used for convenience, since the Hamiltonian is diagonal when $J_{xy} = 0$, allowing to isolate the localized spin dynamics. The realistic case is recovered for $J_z = J_{xy} = J$, for which Eq. (1.13) becomes Eq. (1.12). As for the Schrieffer-Wolff transformation, the bosonization tends to map two models with equivalent low energy properties, with a linearized spectrum around the Fermi energy. Each fermion is first of all separated in a left and right moving parts, corresponding to fermions on each branch of the linearized spectrum: $c_{k\sigma} = c_{k\sigma,R} + c_{k\sigma,L}$, and they are sorted in an odd and even contributions with the following canonical transformation:

$$\begin{cases} A_{k\sigma,o} = \frac{1}{\sqrt{2}}\left(c_{-k\sigma,L} - c_{k\sigma,R}\right) \\ A_{k\sigma,e} = \frac{1}{\sqrt{2}}\left(c_{k\sigma,R} + c_{-k\sigma,L}\right). \end{cases} \tag{1.14}$$

The fermionic commutation relations are still verified by A, that are thus true fermion operators, but that are not local anymore. The total spin density operator on the first site of the chain, $S_z(0)$, is also introduced in the Hamiltonian, which becomes:

$$\begin{aligned}\mathcal{H}_{\text{AKM}} &= \sum_{k,\sigma} \epsilon_k \left(A^\dagger_{k\sigma,e}A_{k\sigma,e} - A^\dagger_{k\sigma,o}A_{k\sigma,o}\right) + \frac{J_z}{2} S_z(0) S^z_d \\ &+ \frac{J_{xy}}{2}\left(\psi^\dagger_{\uparrow,e}(0)\psi_{\downarrow,e}(0) S^-_d + \psi^\dagger_{\downarrow,o}(0)\psi_{\uparrow,o}(0) S^+_d\right).\end{aligned} \tag{1.15}$$

It appears that only even modes A_e couple to the spin. The spin-density operator $S_z(0)$, unlike the fermionic A, obeys bosonic commutation rules, such that the problem is seen as if the density operator was a bosonic excitation $b^\dagger$. To properly define these new bosonic operators, we start to compute the following commutator: $[S_z(0), \mathcal{H}_{\text{AKM}}]$. Then, a new Hamiltonian $\mathcal{H}_{\text{AKM}'}$ is built in terms of $(b^\dagger, b)$ that would produce the same commutation relations $[b, \mathcal{H}'_{\text{AKM}}]$. Once the transformation is done for the density operators, we need to treat single fermion operators $\psi(0)$, and we proceed similarly: we compute its commutator with b, and define a new bosonic operator $\psi \rightarrow e^{ib^\dagger}$ that yields the same commutation relation. One subtility here is that ψ acts in their Fock space by removing a particle to the system, while $b^\dagger$

does not acts on the number of fermionic particles. To be mathematically precise, the so-called Klein factors F are introduced, and they restore the action of the operator on Fock states $\psi \rightarrow Fe^{ib^\dagger}$ [40]. They will appear explicitely at the end of the transformation, in order to simplify next steps.

Once the new bosonic operators have been defined, another canonical transformation introducing operators ϕ and θ is applied:

$$\begin{cases} \phi_\sigma \propto \sum_k \xi_k (b_{k\sigma}^\dagger + b_{k\sigma}) \\ \theta_\sigma \propto \sum_k \xi_k (b_{k\sigma}^\dagger - b_{k\sigma}), \end{cases} \tag{1.16}$$

where ξ_k is performing the mapping to the bosonic Hamiltonian. The real power of these new bosonic operators, is that we can perform a linear combination such that the spin and charge degrees of freedom of the Hamiltonian explicitly separate. The linear combination is the same for ϕ and θ, and reads:

$$\begin{cases} \phi_c = \frac{1}{\sqrt{2}} \left(\phi_\uparrow + \phi_\downarrow\right) \\ \phi_s = \frac{1}{\sqrt{2}} \left(\phi_\uparrow - \phi_\downarrow\right). \end{cases} \tag{1.17}$$

The charge part of the Hamiltonian becomes quadratic in the bosonic field thus does not playing any role in the spin-part dynamics. The charge part is then chosen to be in its vacuum state (equivalently to any state), and only the following spin part remains:

$$H_\sigma = H_\sigma^0 + \frac{J_z S_d^z}{\sqrt{2}\pi} \left(\nabla\theta_s(0) - \nabla\phi_s(0)\right) + \frac{J_{xy}}{2\pi\alpha} \left(S_d^- e^{i\sqrt{2}(\phi_s(0)-\theta_s(0))} + \text{h.c.}\right), \tag{1.18}$$

where α is introduced as a cutoff, to mimic a bandwidth $D \propto 1/\alpha$, so that the mapping is exact for $D \rightarrow \infty$. $\nabla\theta_s(0)$ and $\nabla\phi_s(0)$ are the conjugate fields of $(\theta_s(0), \phi_s(0))$.

The spin and charge degrees of freedom are now separated, but the Hamiltonian that has just been defined is still in a form that can not be processed. The redefinition of the old spin density operators with the new ones as $\psi_\uparrow^\dagger(0)\psi_\downarrow(0) = e^{i\sqrt{2}(\phi_s(0)-\theta_s(0))}$ is close to the form of a single fermionic operator that was defined above $\psi(0) = Fe^{i(\phi_s(0)-\theta_s(0))}$. In fact, there is a factor $\sqrt{2}$ in the exponential that differenciates them. It is known that without the spin flip part, the problem is solved by the unitary transformation $W = e^{i\gamma S_d^z(\phi_s(0)-\theta_s(0))}$, where γ has to be chosen properly. Such a transformation in Hamiltonian (1.18) leads to:

$$WH_\sigma W^\dagger = H_\sigma^{0'} + (2J_z - 2\sqrt{2}\pi\gamma)S_d^z S_z(0) + \frac{J_{xy}}{\pi\alpha} \left(S_d^+ e^{-i(\sqrt{2}-\gamma)(\phi_s-\theta_s)} + \text{h.c.}\right). \tag{1.19}$$

It is obvious that it is interesting to take $\gamma = (\sqrt{2}-1)D$ in order to recover a single fermionic operator, that will greatly simplify Eq. (1.19). It corresponds to the *refermionization* of the bosonic Hamiltonian, in addition to a Jordan-Wigner like transformation which allows to express the spin of the impurity as a spinless fermion:

$$\frac{1}{2}S_d^z = d^\dagger d - \frac{1}{2}, \quad S_d^+ F = d^\dagger, \quad \text{and} \quad c_0 = F e^{i(\phi_s - \theta_s)}. \tag{1.20}$$

Introducing the new fermionic operators in Eq. (1.19) changes the model to the IRLM, which describes the coupling between a spinless impurity level and a Fermi sea:

$$\mathcal{H}_{\text{IRLM}} = \sum_k \epsilon_k c_k^\dagger c_k + V\left(d^\dagger c_0 + \text{h.c.}\right) + U\left(d^\dagger d - \frac{1}{2}\right)\left(c_0^\dagger c_0 - \frac{1}{2}\right). \tag{1.21}$$

The parameters U and V are defined from those of the AKM in the following way:

$$U = \frac{J_z}{\sqrt{2}} - \pi(\sqrt{2} - 1)D, \quad V = \frac{J_{xy}}{2\sqrt{\pi\alpha}}. \tag{1.22}$$

Historically, this mapping was performed to study the Kondo model from the IRLM at $U = 0$, which corresponds to a non-interacting resonant level that can be solved analytically. This point of parameter space is known as the Toulouse point, that was studied before the mapping through bosonization [32], and corresponds to the strongly anisotropic Kondo model at $J_z = \pi\sqrt{2}(\sqrt{2} - 1)D$. Hence, it provides an exact benchmark for different methods in a non-trivial limit, in which any observable in terms of the new fermion d can be computed. The mapping has been done in the spin sector of the Kondo model, such that the charge of the d level in the IRLM reflects that of the original spin, which carries all the low energy physics.

Beyond the Toulouse limit, the strong coupling regime $J_z \to 0$ of the original Kondo model can be investigated through the IRLM when $U \to -\pi(\sqrt{2} - 1)D$. The relevant physics to explore lies at $U < 0$, where the Kondo regime is expected to appear through the scale T_K, that will be defined for this model in the following section. The IRLM itself in the interesting Kondo regime is however not realistic, since it would describe negative Coulomb interaction between electrons on the impurity and in the reservoir. It is therefore only considered for its low energy properties, which are identical to those of the Anderson model. Its spinless nature makes it an appealing model to study numerically, lowering the cost of simulations and simplifying its implementation.

References

1. Bruus H, Flensberg K (2004) Many-body quantum theory in condensed matter physics—an introduction. OUP Oxford
2. Coleman P (2015) Introduction to many-body physics. Cambridge University Press
3. Giamarchi T, Press OU (2004) Quantum physics in one dimension. Clarendon Press. ISBN: 978-0-19- 852500-4
4. Altland A, Simons BD (2010) Condensed matter field theory, 2nd edn. Cambridge University Press
5. Hewson AC (1993) The Kondo problem to heavy fermions. Cambridge University Press
6. Hirsch JE, Fye RM (1986) Monte Carlo method for magnetic impurities in metals. Phys Rev Lett 56:2521–2524
7. Prokof'ev NV, Svistunov BV, Tupitsyn IS (1996) Exact quantum Monte Carlo process for the statistics of discrete systems. J Exp Theor Phys Lett 64:911–916
8. White SR (1992) Density matrix formulation for quantum renormalization groups. Phys Rev Lett 69:2863–2866
9. Wilson KG (1975) The renormalization group: critical phenomena and the Kondo problem. Rev Mod Phys 47:773–840
10. Anderson PWA (1994) Career in theoretical physics. World Scientific
11. Ashcroft N, Mermin N (2011) Solid State Physics. Cengage Learning. ISBN: 978-81-315-0052-1
12. De Haas WJ, van den Berg GJ (1936) The electrical resistance of gold and silver at low temperatures. Physica D: Nonlinear Phenomena 3:440–449
13. Casteljau PDFD, Friedel J (1956) Étude de la résistivité et du pouvoir thermoélectrique des impuretés dissoutes dans les métaux nobles. fr. J de Physique et le Radium 17:27–32. ISSN: 0368-3842, 2777-3442
14. Anderson PW (1961) Localized magnetic states in metals. Phys Rev 124:41–53
15. Kondo J (1964) Resistance minimum in dilute magnetic alloys. Prog Theor Phys 32:37–49
16. Schrieffer JR, Wolff PA (1966) Relation between the Anderson and Kondo Hamiltonians. Phys Rev 149:491–492
17. Anderson PW (1970) A poor man's derivation of scaling laws for the Kondo problem. en. J Phys C: Solid State Phys 3:2436–2441. ISSN: 0022-3719
18. Abrikosov AA, Migdal AA (1970) On the theory of the Kondo effect. en. J Low Temp Phys 3:519–536. ISSN: 1573-7357
19. Fowler M, Zawadowski A (1971) Scaling and the renormalization group in the kondo effect. Solid State Commun 9:471–476. ISSN: 0038-1098
20. Nozières P (1974) A "fermi-liquid" description of the Kondo problem at low temperatures. en. J Low Temp Phys 17:31–42. ISSN: 1573-7357
21. Andrei N, Lowenstein JH (1981) Scales and Scaling in the Kondo Model. Phys Rev Lett 46:356–360
22. Wiegmann PB (1981) Exact solution of the s-d exchange model (Kondo problem). en. J Phys C: Solid State Phys 14:1463–1478. ISSN: 0022-3719 (1981)
23. Andrei N, Furuya K, Lowenstein JH (1983) Solution of the Kondo problem. en. Rev Mod Phys 55:331–402. ISSN: 0034-6861
24. Weymann I, von Delft J, Weichselbaum A (2015) Thermalization and dynamics in the single impurity Anderson model. en. Phys Rev B 92:155435. ISSN: 1098-0121, 1550-235X
25. Rosch A (2012) Wilson chains are not thermal reservoirs. en. The European Phys J B 85:6. ISSN: 1434-6028, 1434-6036
26. Ashoori RC (1996) Electrons in artificial atoms. Nature 379
27. Pustilnik M, Glazman L (2004) Kondo effect in quantum dots. J Phys: Cond Matter 16:R513–R537
28. Cronenwett SM, Oosterkamp TH, Kouwenhoven LP (1998) A Tunable Kondo Effect in Quantum Dots. Science 281:540

29. Pohjola T et al (1997) Resonant tunneling through a two-level dot and double quantum dots. en. Europhys Lett (EPL) 40:189–194. ISSN: 0295-5075, 1286-4854
30. Georges A, Kotliar G, Krauth W, Rozenberg MJ (1996) Dynamical mean-field theory of strongly correlated fermion systems and the limit of infinite dimensions. Rev Mod Phys 68:13–125
31. Kotliar G et al (2006) Electronic structure calculations with dynamical mean-field theory. Rev Mod Phys 78:865–951
32. Toulouse G (1969) Expression exacte de l'énergie de l'état de base de l'hamiltonien de Kondo pour une valeur particulière de Jz C R Acad Sc Paris t 268:1200–1203
33. Blume M, Emery VJ, Luther A (1970) Spin-Boson systems: one-dimensional equivalents and the Kondo problem. Phys Rev Lett 25:450–453
34. Filyov VM, Wiegmann PB (1980) A method for solving the Kondo problem. en. Phys Lett A 76:283–286. ISSN: 0375-9601
35. Wiegmann P, Finkel'Shteĭn A (1978) Resonant-level model in the Kondo problem. Soviet J Exp Theor Phys 102
36. Anderson PW, Yuval G (1969) Exact results in the Kondo problem: equivalence to a classical one-dimensional Coulomb Gas. Phys Rev Lett 23:89–92
37. Guinea F, Hakim V, Muramatsu A (1985) Bosonization of a two-level system with dissipation. Phys Rev B 32:4410–4418
38. Von Delft J, Schoeller H (1998) Bosonization for Beginners. Refermionization for Experts. arXiv:cond-mat/9805275
39. ZaráLand G, von Delft J (2000) Analytical calculation of the finite-size crossover spectrum of the anisotropic two-channel Kondo model. Phys Rev B 61:6918–6933
40. Haldane FDM (1979) Coupling between charge and spin degrees of freedom in the onedimensional Fermi gas with backscattering. en. J Phys C: Solid State Phys 12:4791–4799. ISSN: 0022-3719

Chapter 2
IRLM and Kondo Correlations

In the previous chapter, we traced the history of dilute magnetic impurities in metals and quantum dots. The SIAM and Kondo models were presented, originally introduced to describe the situation. The IRLM, a simpler variant that reports the same low energy properties that are relevant in these models, was also introduced. In this chapter, general equilibrium properties of this model are discussed without and with interactions, and problems we will focus on in the following are introduced.

2.1 Non-interacting Limit $U = 0$

In the non-interacting problem, defined through Eq. (1.21) for $U = 0$, the Hamiltonian is quadratic in the fermion fields. In this regime, there are plenty of ways to solve the problem, depending on the quantity of interest. In this section, two observables are introduced: the occupancy of the impurity, which gives information about the degeneracy of the impurity and its dynamics through Eq. (1.20), and the local density of states of the localized level. The latter is a measurable quantity that shows the spectrum of the states arising from the coupling between the impurity and states of the conduction band. The resonant level describes equivalently the non-interacting Anderson model in each spin sector, which are not coupled without interactions. It is natural to find the Green's functions of the IRLM that are directly related to these observables, and equation of motion (EOM) theory is used to derive them. In the case of non-interacting particles, we can find a closed solution for the EOM, and all quantities can be derived exactly. While interactions are turned on, correlators of higher and higher order are getting involved, and approximations are required to close the problem. The Green's function of the impurity is defined by the following two-time correlator:

$$G_{dd}(\tau) = -\langle \mathcal{T}\big(d(\tau)d^{\dagger}(0)\big)\rangle_0, \tag{2.1}$$

M. Debertolis, *Quantum Impurity Problems in the Framework of Natural Orbitals*, Springer Theses, https://doi.org/10.1007/978-3-031-47233-6_2

where $\mathcal{T}$ is the Dyson's time ordering operator, $\tau = i(t_2 - t_1)$ is an imaginary time, and the subscript 0 means that the average is performed in the ground state (this subscript is implicit and will be omitted in the following). This imaginary time is used for convenience: its Fourier transform is the so-called Matsubara frequency Green's function $G(i\omega_n)$ that has a lot of useful properties that simplifly calculations. Matsubara frequencies are defined for fermions as $\omega_n = (2n+1)\pi/\beta$, where β is the inverse temperature $\beta = 1/T$. The retarded Green's function, that is related to physical quantities, is the analytical continuation of the latter on the real axis, $G^{\mathrm{R}}(\omega) = \lim_{\eta\to 0} G(i\omega_n \to \omega + i\eta)$, so that only use Matsubara frequencies will be used here. The fermionic operators are defined in the Heisenberg picture $d(\tau) = e^{\mathcal{H}\tau} d\, e^{-\mathcal{H}\tau}$. Equation (2.1) is derived with respect to τ:

$$\begin{aligned}\partial_\tau G_{dd}(\tau) &= \partial_\tau \left(\langle d^\dagger(\tau) d(0)\rangle \theta(-\tau) - \langle d(\tau) d^\dagger(0)\rangle \theta(\tau)\right) \\ &= -\{d, d^\dagger\}\delta(\tau) - \langle \mathcal{T}\left([\mathcal{H}, d]_\tau\, d^\dagger(0)\right)\rangle.\end{aligned} \tag{2.2}$$

The commutator with the Hamiltonian being computed, the equation is transformed to Fourier space to perform the derivative, with:

$$G_{dd}(\tau) = \frac{1}{\beta} \sum \omega_n G_{dd}(i\omega_n), \tag{2.3}$$

which simplifies the computation of the derivative as $\partial_\tau G(i\omega_n) = -i\omega_n G(i\omega_n)$:

$$i\omega_n G_{dd}(i\omega_n) = 1 + \epsilon_d G_{dd}(i\omega_n) + V G_{cd}(i\omega_n). \tag{2.4}$$

The mixed Green's function $G_{cd}(i\omega_n) = \sum_k G_{kd}(i\omega_n)$ appears in the equation, and its EOM is also derived using the same method:

$$i\omega_n G_{kd}(i\omega_n) = \epsilon_k G_{kd}(i\omega_n) + V G_{dd}(i\omega_n). \tag{2.5}$$

Hence solving this system for the two Green's functions leads to the following result for $G_{dd}(i\omega_n)$:

$$G_{dd}(i\omega_n) = \frac{1}{i\omega_n - \epsilon_d - \Gamma(i\omega_n)}, \tag{2.6}$$

with $\Gamma(i\omega_n) = \sum_k V^2/(i\omega_n - \epsilon_k)$ the hybridization function, encoding the tunneling properties between the impurity and the bath. The same kind of equations can be derived to find the Green's function of the first site of the bath, which couples to the impurity:

$$G_{cc}(i\omega_n) = \sum_k \frac{1}{i\omega_n - \epsilon_k} + \sum_{kk'} \frac{V^2 G_{dd}(i\omega_n)}{(i\omega_n - \epsilon_k)(i\omega_n - \epsilon_{k'})}. \tag{2.7}$$

The first part corresponds to the free electronic bath, while the second one proportional to V^2 reflects the scattering processes on the impurity. The different Green's functions of this model are closely related to each other, and these results can be recovered in a more compact form with the following field:

$$\psi = \begin{pmatrix} d \\ c \end{pmatrix}, \tag{2.8}$$

whose Green's function is a matrix $\boldsymbol{G}$, from which every Green's function of the problem is recovered:

$$\boldsymbol{G}^{-1}(i\omega_n) = \begin{pmatrix} G_{dd} & G_{dc} \\ G_{cd} & G_{cc} \end{pmatrix}^{-1} = \begin{pmatrix} i\omega_n - \epsilon_d & -V \\ -V & \Gamma(i\omega_n)^{-1} \end{pmatrix}. \tag{2.9}$$

At this point, it is not much work to compute occupancies n_d and n_c, and the corresponding densities of states. For the latter, the propagators are written with Matsubara frequencies, such that they need to be analytically continued to recover the real frequency spectrum. The occupancy at $T = 0$ reads:

$$n_d = 1 - G_{dd}(t = 0^+) = 1 - \int_{-\infty}^{+\infty} \frac{\mathrm{d}\omega}{2\pi} G_{dd}(\omega + i0^+) e^{-i\omega 0^+}, \tag{2.10}$$

where a continuous bath with a linear dispersion $\epsilon_k \propto k$ is considered, with a large bandwidth $D \to +\infty$. Under these assumptions the hybridization $\Gamma(i\omega)$ can be integrated analytically and the usual form of the occupancy [1] is recovered:

$$n_d = \frac{1}{2} - \frac{1}{\pi}\text{atan}\left(\frac{\epsilon_d}{\pi\rho_0 V^2}\right), \tag{2.11}$$

with $\rho_0 = \frac{1}{2D}$ for a constant density of states in the energy range $\epsilon \in [-D, D]$ and 0 elsewhere. The impurity occupancy n_d is plotted on the left panel of Fig. 2.1 for different values of the coupling V. As expected, as ϵ_d goes from negative to positive values, the dot goes from almost occupied to almost empty. As V is quenched to 0, the impurity decouples more and more and becomes thus easily polarized, while as V gets larger it hybridizes with more and more states, such that shifting the level with $|\epsilon_d| \gtrsim 0$ only slightly changes the occupation. This effect is similar to a susceptibility in an external field ϵ_d controlling the width of the crossover between $n_d = 1$ and $n_d = 0$. At $\epsilon_d = 0$ the Hamiltonian is particle-hole (PH) symmetric and this property leads to an exactly half-occupation of the impurity $n_d = 1/2$, where fluctuations are maximum.

These effects of hybridization are clearly seen in the LDOS of the impurity. This is a dynamic observable that is usually defined through the retarded Green's function:

$$\rho_d(\omega) = -\frac{1}{\pi}\mathrm{Im}\left(G^{\mathrm{R}}_{dd}(\omega)\right) = \frac{\Gamma_0/\pi}{(\omega-\epsilon_d)^2+\Gamma_0^2}. \tag{2.12}$$

Numerically, this quantity is defined equivalently with the Lehmann representation of the Green's function, that uses the completeness of the Hamiltonian eigenvectors written as the identity $\mathbb{1} = \sum_\alpha |\alpha\rangle\langle\alpha|$:

$$\rho_d(\omega) = \frac{1}{\mathcal{Z}}\sum_\alpha |\langle\alpha|d^\dagger|0\rangle|^2\delta\left(\omega-(E_\alpha-E_0)\right), \tag{2.13}$$

where $\mathcal{Z}$ is a norm, 0 labels the state of lowest energy, state that reads $\prod_{E_i<0}\xi_i^\dagger|0\rangle$, if the set of orbitals ξ_i diagonalizes $\mathcal{H}$. Since the number of modes is finite in numerical simulations, the δ distributions are smoothed by Lorentzian functions of width $\eta < \delta E$, where δE is the energy spacing. The spectral functions of the impurity and of the first site of the bath are reported in the right panel of Fig. 2.1: the impurity hybridizes with states of the bath over an energy scale $\Gamma_0 = \pi V^2/2D$, which are depleted from the bath as can be seen in ρ_c. This function is also closely related to the occupancy [2] n_d as:

$$n_d = \int_{-\infty}^{+\infty}\frac{\mathrm{d}\omega}{2\pi}\rho_d(\omega)n_F(\omega). \tag{2.14}$$

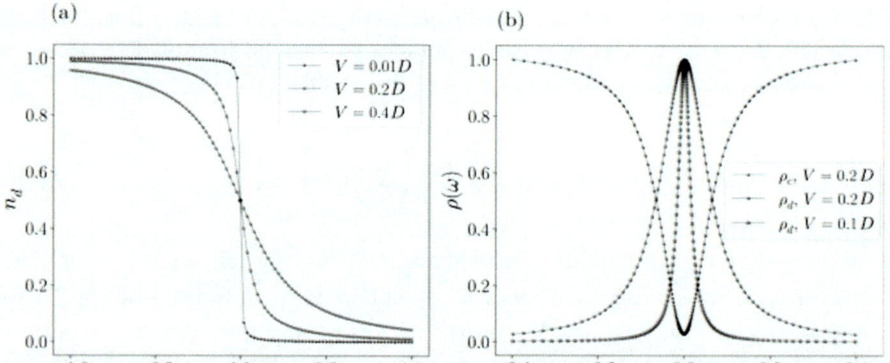

Fig. 2.1 **a** Impurity level occupancy n_d for different values of V computed numerically with a discrete version of Eq. (2.14) using the Lehmann representation against ϵ_d. The lines are guides for the eye. **b** Density of states of the impurity and of the first site of the bath. The points are taken on a logarithmic grid to gain accuracy at low energies, where the peak is narrow for small V. Each curve is divided by its maximum value for readability, changing the norm of each function to $\int_{-\infty}^{\infty}\mathrm{d}\omega\rho(\omega) = \max(\rho(\omega))^{-1}$

Here, $n_F(\omega)$ is the temperature dependent Fermi occupation function, and we take it here at zero temperature (that is a step function up to the Fermi energy) to find back the result (2.11). All these results are relatively simple to derive in the non-interacting limit, but going to the fully interacting problem is a whole different story. At small $|U| \ll D$, perturbation theory can provide quantitative results, but pushing U to larger values requires the use of more powerful methods, which are examined now.

2.2 Numerical Renormalization Group

In order to investigate the physics deeper in the strong coupling regime, interactions have to be fully taken into account. In the framework of quantum impurities, despite the large number of available methods that we briefly discussed above, the numerical renormalization group (NRG) proposed by Wilson [3] is the most practical non-perturbative method to solve the Kondo problem at zero or finite temperature [4, 5]. The algorithm works for any form of the impurity, whose only requirement is to be diagonalized exactly, coupled to a non-interacting bath of fermions or bosons. The information of the coupling between the environment and the bath is encoded in the hybridization function $\Gamma(\omega)$, that will be taken constant in the calculations in the interval $[-D, D]$.

To accurately describe the low energy physics of quantum impurity problems, the NRG relies on a transformation of the bath degrees of freedom. Using the well separated energy scales of these problem, from D to T_K, the density of states of the bath is discretized logarithmically. This is equivalent to a discretization of $\Gamma(\omega)$, such that the impurity couples to states exponentially close to the Fermi level, that would require exponentially large systems without this transformation. Then, the problem is mapped onto a semi-infinite chain, which is constructed iteratively. This iterative procedure involves successive transformations of the Hamiltonian parameters and constitutes the renormalization group flow of parameters and of any observable that is calculated.

Along the flow, the Fock space dimension is growing exponentially, and high energy states are discarded at each iteration to keep this dimension fixed. We will not enter in every details here that are not needed for comprehension, and the reader can refer to Ref. [6] for a complete review. The detailed NRG implementation of IRLM and SIAM is given in Appendices A and B respectively.

2.2.1 Writing the Hamiltonian on a Logarithmic Grid

Before performing the RG procedure, the Hamiltonian with a continuous conduction band must be rearranged in the so-called Wilson chain. This logarithmic discretiza-

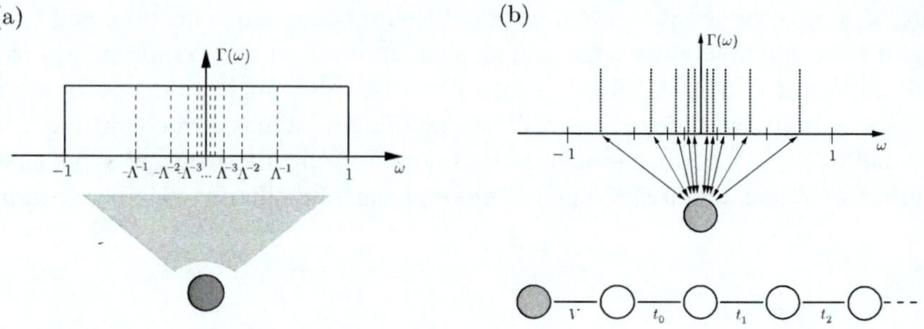

Fig. 2.2 The figures are taken from Ref. [6]. **a** Coupling of the impurity (green dot) to a bath via the continuous hybridization $\Gamma(\omega)$. **b** Discretization of $\Gamma(\omega)$ on the logarithmic grid, and mapping to a 1D chain with hopping parameters t_n

tion is parametrized by the unique parameter Λ, such that one state of the band is kept in each continuous energy interval $[\omega_{n+1}, \omega_n]$, $\omega_n = \Lambda^{-n}$, with $n = 0, \ldots, \infty$. $\Lambda \to 1$ corresponds to the continuous limit of the discretized model, and usually Λ takes values around 2, but can be taken as large as required for particular cases. However, the larger the Λ, the worse becomes the description of high energy properties of order D.

Operators of the bath are redefined to fit on the grid by some change of basis, but the mathematical details are skipped here. Once the Hamiltonian is reshaped, it is tridiagonalized through a series of transformations of operators. Figure 2.2 sketches the mapping. The principal information of this tridiagonalization is that the impurity part did not change, and that hopping terms along the chain to which it is coupled are decreasing exponentially. The Hamiltonian (1.21) is transformed to:

$$\mathcal{H}_{\text{IRLM}} = \epsilon_d d^\dagger d + V\,(d^\dagger c_0 + \text{h.c.}) + U\left(d^\dagger d - \frac{1}{2}\right)\left(c_0^\dagger c_0 - \frac{1}{2}\right) + \sum_{n=0}^{\infty} t_n\,(c_{n+1}^\dagger c_n + \text{h.c}). \tag{2.15}$$

For a general hybridization $\Gamma(\omega)$, the hopping terms t_n have no explicit form and are defined through a recursive series $t_{n+1} = u(t_n)$. However, if $\Gamma(\omega)$ is taken constant, which is accurate enough for the low energy properties that will be considered, a closed form exists:

$$t_n = \frac{(1+\Lambda^{-1})(1-\Lambda^{-n-1})}{2\sqrt{(1-\Lambda^{-2n-1})(1-\Lambda^{-2n-3})}}\Lambda^{-n/2}. \tag{2.16}$$

Hopping terms along the chain only depend on the discretization parameter Λ, and $t_n \sim \Lambda^{-n/2}$ as $n \to \infty$. In actual simulations, the chain is cut at a given site $n = N$, so that the smallest eigenenergies of the Hamiltonian are of order $\Lambda^{-N/2}$. Reaching the thermodynamic limit in the strict sense requires to deal with infinite systems, but in this setup choosing N such that the eigenenergies are decades below the lowest energy scale (for quantum impurity problems it is usually T_K) meets our needs. To reach lower energies for a same N, larger Λ should be taken, but this has an impact on the states under consideration. As an example, taking $\Lambda = 2$ and $N = 50$ sites already leads to energies (in units of $D = 1$) of order 10^{-8}, that is more than enough for realistic situations in which $T_K \sim 10^{-3}$.

Before going further with the iterative diagonalization of the chain, two details must be clarified about the precedent transformation. First, the mapping to the chain does not require the original problem to be 1D, and can be equivalently performed for baths of higher dimension. Second, the mapping presented above is not exact as was the bosonization. Besides the discretization, some couplings between the impurity with orbitals defined on each bath state n have been thrown away. Nevertheless, these couplings correspond to subdominant processes, and it can be verified that they do not play any tangible role in simulations.

2.2.2 *Iterative Diagonalization of the Chain*

Once the problem has been expressed in a convenient way into Eq. (2.15), it remains to solve it. To do so, the chain Hamiltonian will be constructed through a series of transformations that approaches $\mathcal{H}$ in the limit $n \to \infty$:

$$\mathcal{H} = \lim_{n\to\infty} \Lambda^{-(n-1)/2} H_n. \tag{2.17}$$

H_n is the intermediate Hamiltonian at step n of the RG procedure, and the factor $\Lambda^{-(n-1)/2}$ cancels the n dependence of the hopping terms t_n, in such a way that eigenenergies of H_n are always of order 1 to avoid numerical instabilities. The correct spectrum can be recovered easily by multiplying it by the inverse factor. The starting point of this algorithm, H_0, is a part of the Hamiltonian (2.15) that can be diagonalized exactly. Generally, only the impurity part is considered, but as IRLM interactions involve also the first site of the chain, it is convenient to consider the impurity and this first site as a starting point. Then, it is diagonalized in Fock space, and the space is augmented by adding to the system the second site of the chain, and so on. Unlike usual RG transformations, the flow does not really renormalize a fixed set of parameters of the Hamiltonian but rather the full Hamiltonian itself, the number of couplings increasing over successive transformations. It is suitable here to characterize the flow with the many-body energy spectrum, that is defined at each step as:

$$H_n|r\rangle_n = E_n^r|r\rangle_n, \tag{2.18}$$

with $|r\rangle_n$ the complete set of eigenvectors at step n. To build H_{n+1}, the degrees of freedom $|s\rangle$ of the new site are added to the current Hamiltonian via the tensor product:

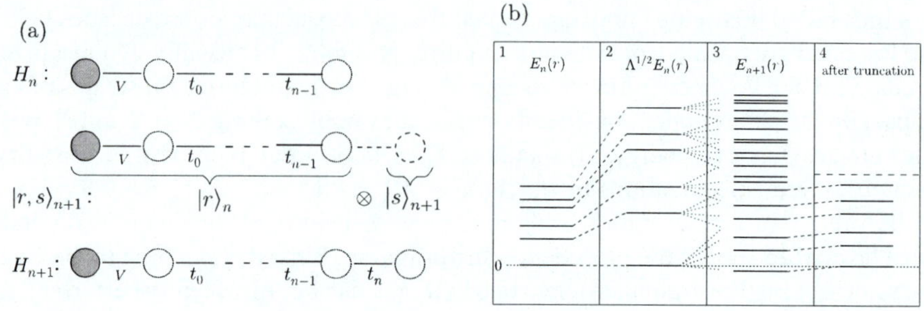

Fig. 2.3 The figures are taken from Ref. [6]. **a** Sketch of the iterative construction of the Hamiltonian. Degrees of freedom $|s\rangle_{n+1}$ are added to the diagonal part of basis vectors $|r\rangle_n$. **b** Truncation scheme when the number of states exceeds a given number N_{kept}. 1→ 2: rescaling of the eigenvalues to be of order 1. 2→ 3: extension of the Fock space by the added degrees of freedom. 3→ 4: shift of the eigenvalues to have $E_0 = 0$ and truncation of high energy states

$$|r'\rangle_{n+1} = |r\rangle_n \otimes |s\rangle_{n+1}, \quad H_{n+1} = \Lambda^{1/2} \sum_r E_n^r |r'\rangle_{n+1} \langle r'| + \Lambda^{n/2} t_n (c_{n+1}^\dagger c_n + \text{h.c.}), \tag{2.19}$$

and in the present case, the new site can have two possible states $|s\rangle$ (empty or full) such that the size of the total space is doubled at each step. c_{n+1} and c_n are expressed in the new basis $|r'\rangle_{n+1}$, with the former acting on $|s\rangle_{n+1}$ and the latter on $|r\rangle_n$. Then the new Hamiltonian is diagonalized, and the algorithm is iterated until the desired convergence is reached, i.e. $n = N$. The procedure is depicted in the left panel of Fig. 2.3.

However, the Fock space dimension is growing exponentially, and in a few steps the reached dimension can not be handled by a computer. Therefore, the Fock space is truncated at each iteration: only the N_{kept} lowest energy states are kept while the rest are discarded. This truncation is well controlled by the logarithmic discretization, since high energy states of a given iteration are well separated from the low energy ones of the next step. N_{kept} must be chosen as a function of the complexity of the problem and of the desired precision. In general, if Λ is closer to 1, N_{kept} is taken bigger to compensate the thiner separation of sectors. The right panel of Fig. 2.3 shows this truncation procedure. Between the third and fourth steps shown in this figure, the energies have been shifted by E_n^0, that is kept at $E_n^0 = 0$ during the flow.

2.2.3 *Calculating Observables*

We first discuss the case of static observables, which are frequency independent. It is not possible to wait until the last iteration to use the wave function and define an observable. Indeed, the basis is not complete, there is no simple correspondence between the degrees of freedom of the beginning and the final eigenvectors, unless by storing all consecutive transformations, that would be needless work. Though, it is possible to define an operator at the beginning of the flow (or when the corresponding operator is introduced into the flow) as was done for H_0, and follow the successive transformation by projecting this operator on the current basis. At the end, the operator is evaluated in the ground state to obtain the desired static observable. Any observable involving the impurity only can be defined, but the use of the Wilson chain puts strong limitations for quantities that are local in the bath.

On the other hand, for dynamic observables the story is more involved. Along the flow, the algorithm generates successive sets of eigenvectors, making thus natural the use of the Lehmann representation defined before in Eq. (2.13), that can be written for any frequency dependent correlator. After accessing the spectral part, that corresponds to the imaginary part of the observables, the real part is obtained via Kramers–Kronig relations, that will be introduced in a future section. The states taken in the formula will be the ones defined at each step $|r\rangle_n$. As the states are exponentially low in energy, the low frequency properties will be accurately described. However, for higher frequencies, only few states are accessible and everything of the order of D will suffer from the discretization [7–9].

Another problem with taking every states in each iteration is that the completeness is ill defined: some low energy states of step n are decomposed on a set of other low energy states of further iteration, thus counting more than once the same information. This naive use of the Lehmann representation is unappropriate, and methods have been developed to properly define dynamic observables: the DM-NRG [10–13] is an extension that uses the full density matrix to define correctly a complete set of states from the first discarded state to the last kept state. The reader can refer to the cited articles explaining extensively the optimized algorithm. Although powerful NRG algorithms are available to compute such observables, numerical accuracy and the Wilson chain remain the great barriers of this approach and research is still active on finding new ways to improve these calculations.

2.3 Quantum Phase Transition in the IRLM

As expected through the bosonization transformation, the IRLM hosts a quantum phase transition (appearing at zero temperature, and driven by parameters of the system) similar to the ferromagnetic—antiferromagnetic one in the Kondo model. From the mapping between models, the transformation is expected to appear at a negative value of the interaction $U_c \simeq -\pi(\sqrt{2}-1)D$. In the limit $U \to -\infty$, the ground state is doubly degenerate at particle hole symmetry ($\epsilon_d = 0$), with states $d^\dagger c_0^\dagger|0\rangle$ and $|0\rangle$. These states are not coupled by the hybridization V and remain degenerate at large and negative U. The occupation of the impurity in these states is either full or empty, which corresponds to a spontaneous breaking of the particle-hole symmetry under which the occupation should be $1/2$. Through the Jordan-Wigner transformation, it corresponds to a ground state with a residual magnetic moment, the degeneracy being between $|\uparrow\rangle$ and $|\downarrow\rangle$, which corresponds to the ferromagnetic phase. As the ground states of the two phases are not adiabatically connected, it can be expected that the occupation n_d is discontinuous at the transition.

The Kondo correlated regime lies in the symmetric phase $0 > U > U_c$, in which the Kondo temperature T_K vanishes as we approach the transition point U_c. In the IRLM, T_K is calculated through the charge susceptibility [1, 6, 14] of the impurity to a local *field* ϵ_d, at PH symmetry:

$$T_K = \frac{1}{4\chi}, \quad \chi = \left.\frac{\partial n_d}{\partial \epsilon_d}\right|_{\epsilon_d=0}. \tag{2.20}$$

In practice, n_d is calculated with two NRG simulations at $\epsilon_d = 0$ and $\epsilon_d = \eta$, with $\eta \ll T_K$ in order not to introduce an infrared cutoff. T_K is also recovered from the flow as the crossover scale to the Fermi-liquid fixed point in which the impurity is screened, that can be used to check if Eq. (2.20) gives the correct order of magnitude. As T_K becomes smaller, the number of kept states of the NRG and the size of the chain N are increased for the thermodynamic limit to be reached. While T_K becomes of the order of the numerical accuracy $\simeq 10^{-15}D$, there is no hope to get accurate results, the impurity being polarized by the numerical noise, such that an exponential fit of T_K in the range $\left[10^{-14}, 10^{-6}\right]$ is used to find the transition point U_c. The results are plotted in the left panel of Fig. 2.4, and the fit predicts $U_c = -1.311D$, that corresponds to the expected value $-\pi(\sqrt{2}-1) \simeq -1.301D$ at small V.

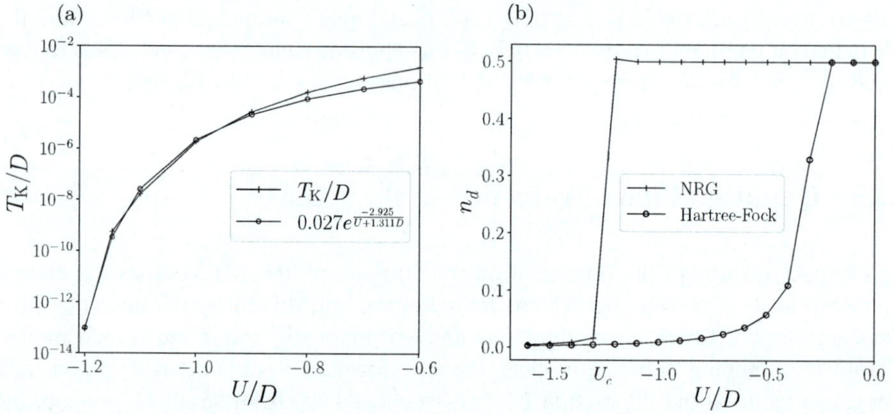

Fig. 2.4 **a** Kondo temperature of the IRLM computed in NRG through Eq. (2.20). An exponential fit for $U \to U_c$ is added to extrapolate T_K closer to the transition. **b** Occupation of the impurity for $\epsilon_d = 0$ and $V = 0.15$ ($\Gamma \simeq 3.510^{-3}$) calculated in NRG and with the Hartree–Fock approximation. NRG simulations are performed with $\Lambda = 2$, $N = 110$ sites and with $N_{\text{kept}} \simeq 600$ kept states for $U > -1.0$, and $N_{\text{kept}} \simeq 1300$ for $U_c < U < -1.0$. The lowest eigenenergies are of order $2^{-55} \simeq 10^{-17}$, so that convergence is assured

In both phases, the occupation of the localized level which characterizes the transition is also easily computed in NRG. In Fig. 2.4, NRG results are contrasted with the Hartree–Fock solution of the IRLM to show the need for such a method. The Hartree–Fock approximation consists in neglecting the fluctuations of the operators participating to the interaction. The interaction term becomes quadratic in the fermionic fields, and its effect is the renormalization of the parameters involving the impurity:

$$\tilde{\epsilon}_d = \epsilon_d + U\langle c_0^\dagger c_0 \rangle, \quad \tilde{V} = V - U\langle d^\dagger c_0 \rangle, \quad \epsilon_0 = U\langle d^\dagger d \rangle. \tag{2.21}$$

ϵ_0 is an interaction-induced potential on the first site that reads $\epsilon_0 c_0^\dagger c_0$ in the Hamiltonian. The optimized renormalization of parameters is found self-consistently: equations for the free propagators are solved numerically by diagonalizing an effective Hamitonian with the renormalized parameters until convergence. While this method provides correct results for small $|U|$, where fluctuations are small, it becomes totally false in the correlated regime $|U| \gg \Gamma$, such that perturbation theory would require the resummation of diagrams at all orders. Within Hartree–Fock calculation, n_d is continuous at the transition and the critical U_c is under-estimated because of the early polarization of the impurity by effective potentials. The correct behavior is accurately reproduced by NRG simulations, but precision is lost due to the quench of T_K when U becomes too close to U_c.

While studying the strongly correlated regime, it should be clear that more and more computational effort should be put when $U \to U_c$ since fluctuations are maximal, but physical systems have intrinsic limitations (temperature, finite size) that prevent T_K to be as small as the ones considered here. The quantum phase transition of the IRLM will be investigated again in Chap. 3 in the framework of natural orbitals, that will be defined therein.

2.4 The Kondo Screening Cloud

Using scaling arguments, it was explained in Chap. 1 that the spin of the Kondo impurity (or charge of IRLM) was quenched by conduction electrons surrounding it. However, the spatial mechanism responsible for such a screening has not been discussed yet. An incorrect picture would be that the two electrons with opposite spins in the impurity forms a spin zero state and thus decouple from the host. However, the price to pay is the onsite Coulomb interaction, and the previous scaling arguments shown that the interaction renormalizes to large values in the scaling limit, preventing this kind of local singlet. Thus, the formation of a *compensation cloud*, that is well-known today as the *Kondo cloud* or *screening cloud*, involving electrons of the host has been adopted to explain the formation of the singlet [15–18]. While the temperature is lowered, more and more electrons participate to compensate the local moment, and once the scale T_K is exceeded, the screening process is complete and the Fermi-liquid fixed point is reached. The energy of electrons participating in the process ranges from D down to $\sim T_K$, which corresponds to a spatial extension up to the scale $L_K \simeq 1/T_K$. As T_K is exponentially low, L_K is expected to be large, up to some microns in real setups.

The long range nature of the screening mechanism, including a wide range of energy scales makes it difficult to study, and a lot of theoretical works were done to provide a complete description of the phenomenon [19–42]. The usual observable to study this screening cloud is the equal-time spin-spin correlator of the impurity and conduction electrons. This quantity measures correlations spreading from the impurity to the bath, and reads for the Kondo model:

$$\chi_{t=0}(x) = \langle \mathbf{S}_d \cdot \mathbf{S}(x) \rangle, \tag{2.22}$$

where the bath can be continuous or discrete: $\chi_i = \langle \mathbf{S}_d \cdot \mathbf{S}_i \rangle$. As usual, spin correlations in the Kondo model induces analogous charge Kondo correlation in the IRLM through the density-density correlator:

$$C_i = \langle \left(d^\dagger d - \langle d^\dagger d \rangle\right) \left(c_i^\dagger c_i - \langle c_i^\dagger c_i \rangle\right) \rangle. \tag{2.23}$$

The average charges are substracted to reveal the long-distance fluctuations originating from the non-trivial correlations between the impurity and the bath. At PH symmetry with $U > U_c$, the averages are known to be equal to $1/2$. This observable is known to oscillate fast up to large distances, and correlations are expected to be small as the singlet ground states imposes the sum rule:

$$\text{for the Kondo model:} \quad \sum_{i=0}^{N} \chi_i = -\frac{3}{4}, \quad \text{and for the IRLM:} \quad \sum_{i=-1}^{N} C_i = 0, \tag{2.24}$$

where $C_{-1} = \langle d^\dagger d \rangle \langle d^\dagger d \rangle = 1/4$, and it is clear that substracting the averages in Eq. (2.23) imposes the sum of the observable to be zero. These sum rules can also be useful for numerical simulations to investigate the accuracy of the constructed observable.

The correlator oscillates at a wavevector $2k_F$, with inequivalent odd and even components along the discrete chain. At intermediate distances $i \ll L_K$, perturbation theory predicts that the envelope for both components decays as the inverse of the distance to the impurity i^{-1} [43]. In the opposite limit $i \gg L_K$, Fermi liquid arguments predict an i^{-2} decay for the largest component and an i^{-4} decay for the other one [16]. The full crossover at scale $\sim L_K$ is not predicted by such arguments, and neither Bethe ansatz nor NRG (in its simple implementation) can be used to compute accurately such a real space observable. At high temperatures, spin-spin correlations behaves as the inter-impurity Ruderman–Kittel–Kasuya–Yosida (RKKY) coupling in bulk systems [19], that extends the importance of studying the screening cloud around magnetic impurities for bulk materials.

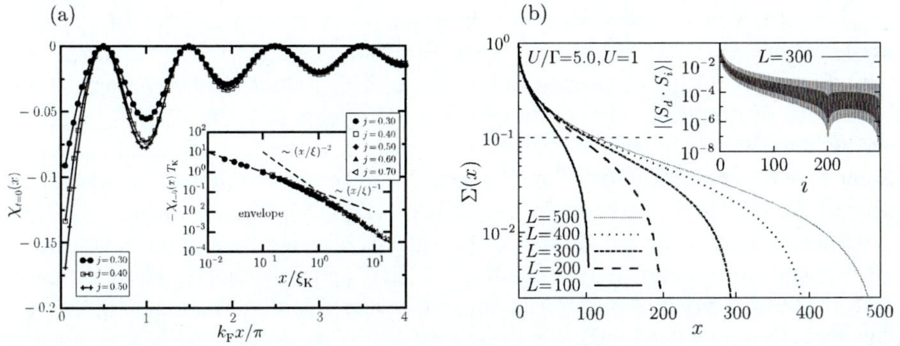

Fig. 2.5 **a** The plot is taken from Ref. [26]. NRG simulation of the spin-spin equal-time correlator $\chi_{t=0}(x)$ in the bath at distance x from the impurity. The inset shows the envelope of largest component of the oscillations of $\chi(x)$, the distance being normalized by the Kondo length L_K, that corresponds to the crossover length between the two different power law behavior. **b** The plot is taken from Ref. [29]. DMRG results of $\Sigma(x)$, the sum of the cloud amplitude up to distance x, $\Sigma(x) = 1 + \sum_{i=0}^{x} \chi_i / \langle S_d \cdot S_d \rangle$, that is plotted for different system size L. The inset shows the spatially resolved observable χ_i for a size of $L = 300$ sites

Beyond the perturbative or scaling arguments, numerical simulations were designed to calculate the cloud accurately in real space. The first advance in that path was a tour de force from Borda [26], who managed to adapt the NRG to obtain real space resolution of observables, circumventing the limitations of the Wilson chain. It uses the combination of the impurity field with one of the bath at distance x in the bath, where one wants to evaluate the observable. In terms of this new field, an observable including the impurity and the site x of the bath can benefit from the accuracy of usual local NRG simulations. However, there is a price to pay: first, the new field brings more complexity to the impurity part, that will be numerically more expensive to converge. Second, the density of states of the effective bath will be energy dependent, such that the hoppings are not given by Eq. (2.16), but by recursive relations that may be unstable numerically. Last, each evaluation of the observable at distance x require an entire NRG simulation. Nevertheless, the results shown in the left panel Fig. 2.5 were a breakthrough for NRG simulations and the understanding of Kondo correlations.

In parallel, Density matrix renormalization group (DMRG) implementations of the SIAM [25, 29] allowed simpler studies of the screening process in different conditions, since the full many-body wave function allows one to compute χ_i in a single shot. Still, in the Kondo regime, the cost of DMRG simulations grows fast (mediating by the bond dimension), and the system sizes are limited to few hundreds of sites, that is not enough to converge observables if T_K is below $10^{-2} - 10^{-3}$. Results from [29] are reported in the right panel of Fig. 2.5, and show the distance at which the impurity gets screened for different system size L. Finite size effects are strong because the thermodynamic limit is not reached, and the sum rule Eq. (2.24) imposes correlations to vanish fast at the boundary for small systems.

Thus, the Kondo screening is well understood on the theoretical side, and the existence of a macroscopic screening phenomenon is considered to be correct. On the other hand, a direct experimental measure of this cloud is still missing. There were a lot of proposals for experiments over the years [25, 28, 45–55], but measuring a non-local observable over such distances is a real technological challenge. Some experiments reported different signatures of Kondo correlations at large length scales [56–58], but the first convincing study was done by Borzenets et al. [44]. Their experiment shows how the Kondo temperature is affected by a perturbation in the electronic channel coupled to a quantum dot. They applied voltage gates at distances L from the impurity, and through a temperature dependent measurement of the conductance, they extracted the corresponding $T_{K,max}$. These gates are at $L = 1.4\,\mu m$, $L = 3.6\,\mu m$ and $L = 6.1\,\mu m$ from the impurity, and applying a voltage in one of them cuts abruptly correlations between the impurity and conduction electrons, if they exists at such a distance. They observed strong oscillations in the measured $T_{K,max}$ when turning on the first gates, and only little changes for the farthest ones. This observation implies that between the two gates must be the true (i.e. without the perturbation by the gates) correlation length of the Kondo state, that is thus of the order of few microns, as predicted. Although this proves the existence of long range correlations around quantum dots, it is still not a direct measurement of spin-spin correlations Eq. (2.22) at all scales, that would definitely close the debate.

2.5 Kondo Correlations in a Disordered Host

Experimental realizations of condensed matter systems are inherently subject to structural disorder, that can strongly affect their properties. In situations studied above, disorder was supposed to be irrelevant, and only clean systems were considered. The effects measured experimentally were correctly described by disorder-free models, confirming the validity of this assumption. However, later experiments in Kondo alloys presented exotic non-Fermi liquid (NFL) behavior at low temperatures [59–64], going against the picture of a quenched moment by conduction electrons. The Fermi-liquid behavior is usually verified by temperature dependent measurements of the resistivity, that should vary as T^2 at low T. Magnetic susceptibilities or specific heat are also commonly measured in these experiments and show a quenching of impurity local moments. Such a NFL behavior of the resistivity ρ is shown in Fig. 2.6 for the Kondo alloy $Y_{1-x}U_xPd_3$ [59]. At high temperature, logarithmic increase of ρ with the temperature is measured as expected by the diffusion on magnetic impurities. However, instead of a saturation at low temperature $\propto T^2$, a linear dependence is measured.

The mechanism at play for in these systems was the presence of non-negligible disorder that triggers a wide range of local T_K, so that a fraction of local moments remain unscreened at any temperature, which was confirmed theoretically [65–69], corroborated by muon spin rotation experiments [60]. To investigate the effect of disorder in Kondo systems, the distribution of Kondo temperatures was calculated. An

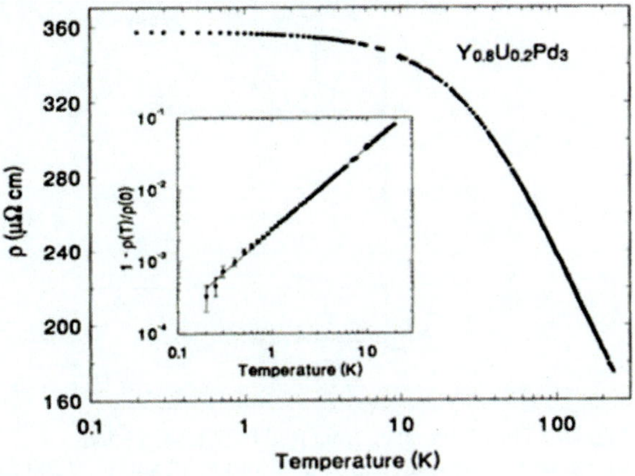

Fig. 2.6 The plot is taken from Ref. [59]. Resistivity ρ against temperature T for a concentration $x = 0.2$. The Kondo—NFL transition is seen around $T = 80$ K. The inset shows the power-law behavior below 20 K, fitted by: $\rho/\rho(0) = 1 - (T/T_0)^{1.13}$, with $\rho(0) = 357.7\,\mu\Omega$ cm and $T_0 = 180$ K. This power-law is the evidence for the NFL behavior

NRG or DMRG study of spatially distributed disorder in a Kondo-correlated regime would be too expensive, so that the full numerical many-body study of these systems was not considered. However, the Kondo temperature of a system can be obtained at a mean field level, requiring only the one-body spectrum of the Hamiltonian. This formula was introduced by Nagaoka in the early times of the Kondo effect [70], and T_K can be computed from:

$$\frac{2D}{J} = \sum_{n=1}^{N} \frac{x_n}{s_n} \tanh\left(\frac{s_n \Delta}{2T_K}\right), \tag{2.25}$$

where $x_n = N|\psi_n(0)|^2$ is the weight of the wave function on the impurity, $s_n = (E_n - E_F)/\Delta$ and Δ the mean level spacing. The distribution of Kondo temperatures $P(T_K)$ computed in Ref. [67] are reported in Fig. 2.7. It appears that in the Kondo regime, when the disorder is increased, the Kondo temperatures become smaller and smaller. Actually, a finite fraction of the T_K is zero, and as the disorder is increased this part becomes dominant. A T_K going to zero means that the impurity is never screened by its surroundings, in such a way that the local moment stays free even at low temperatures.

In Ref. [68], the same low temperature behavior is found, and the fraction of unscreened moments as a function of the interaction J is computed for different disorder strengths. Indeed, the relevant parameter is J/W (W the disorder strength), and $P(T_K = 0) \to 1$ as $J/W \to 0$. The mechanism behind the destruction of the coherent screening by disorder is as follows: there is a disorder induced gap in the density of states of the impurity, that should be compared to the hybridization between the impurity and the bath. As J goes to 0, the hybridization involves states in an energy window of width $\sim T_K$ around the Fermi energy, and the depletion of these states by the disorder is enough to isolate the impurity from its surroundings. Hence, it stays unscreened, and the Fermi liquid fixed point can not be reached. That

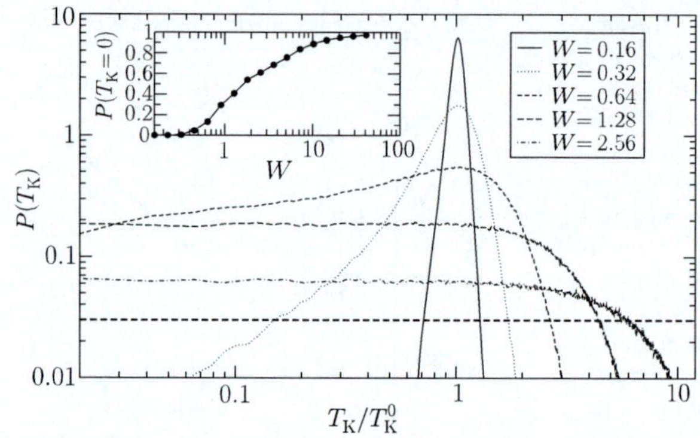

Fig. 2.7 The plot is taken from Ref. [67]. Distribution of Kondo temperatures for different disorder strengths, for a one-dimensional disordered bath, for 1000 realizations of the potential in a chain of 1000 sites, with $J = D/4$. The inset shows the fraction of unscreened impurity at $T = 0$ as a function of the disorder strength W

leads to a divergent magnetic susceptibility as $T \to 0$ that is consistent with NFL experiments. Nonetheless, these studies are limited to the mean field description of T_K, and it would be interesting to see how does the screening cloud disappear when the disorder gets stronger. It would require to compute the observable (2.22) for each realization of the disorder, a challenge that will be addressed in this thesis.

Among the theoretical work concerning Kondo correlations within disordered metallic hosts [65–68, 71–84], disorder in the IRLM has not been studied. In this spinless model, disorder can only be considered in the charge sector, that corresponds to static magnetic disorder in the Kondo model, which is less common than charge disorder. This study will be presented in Chap. 5, and will go further than a mean field calculation of Kondo temperatures. The algorithm presented in Chap. 4 allows a microscopic study of each realization of disorder and thus provides interesting details about spatial correlations spreading in the disordered host. The study will be extended to a Hamiltonian model of disordered quantum impurity using random matrices, presented in Chap. 6. Random matrix theory (RMT) for physical systems has been widely developed over the last century, and its applications to quantum impurities have already proven successful [68, 69, 77, 85, 86]. The presented study will go further by introducing the probability density of the wave function, that allows to compute any local observable concerning the impurity.

References

1. Hewson AC (1993) The Kondo problem to heavy fermions. Cambridge University Press
2. Bruus H, Flensberg K (2004) Many-body quantum theory in condensed matter physics - an introduction (OUP Oxford, 2004)

3. Wilson KG (1975) The renormalization group: critical phenomena and the Kondo problem. Rev Modern Phys 47:773–840
4. Costi TA, Hewson AC, Zlatic V (1994) Transport coefficients of the Anderson model via the numerical renormalization group. J Phys: Condens Matter 6:2519–2558
5. Bulla R, Hewson AC, Pruschke T (1998) Numerical renormalization group calculations for the self-energy of the impurity Anderson model. J Phys: Condens Matter 10:8365–8380. ISSN: 0953-8984
6. Bulla R, Costi TA, Pruschke T (2008) Numerical renormalization group method for quantum impurity systems. Rev Mod Phys 80:56
7. Oliveira WC, Oliveira LN (1994) Generalized numerical renormalization-group method to calculate the thermodynamical properties of impurities in metals. Phys Rev B 49:11986–11994. ISSN: 0163-1829, 1095-3795
8. Freyn A, Florens S (2009) Optimal broadening of finite energy spectra in the numerical renormalization group: application to dissipative dynamics in two-level systems. Phys Rev B 79:121102. ISSN: 1098-0121, 1550-235X
9. Zitko R, Pruschke T (2009) Energy resolution and discretization artifacts in the numerical renormalization group. Phys Rev B 79:085106. ISSN: 1098-0121, 1550-235X
10. Hofstetter W (2000) Generalized numerical renormalization group for dynamical quantities. Phys Rev Lett 85:4
11. Peters R, Pruschke T, Anders FB (2006) A numerical renormalization group approach to Green's functions for quantum impurity models. Phys Rev B 74:245114. ISSN: 1098-0121, 1550-235X
12. Toth AI, Moca CP, Legeza O, Zarand G (2008) Density matrix numerical renormalization group for non-Abelian symmetries. Physical Review B 78:245109. ISSN: 1098-0121, 1550-235X
13. Weichselbaum A, von Delft J (2007) Sum-rule conserving spectral functions from the numerical renormalization group. Phys Rev Lett 99:076402. ISSN: 0031-9007, 1079-7114
14. Ishii H, Yosida K (1967) Zero-temperature susceptibility of a localized spin exchange coupled with the conduction electrons. Progress Theor Phys 38:61–71. ISSN: 0033-068X
15. Müller-Hartmann, E. Spin correlation in dilute magnetic alloys. en. Zeitschrift für Physik A Hadrons and nuclei 223, 277-288. issn: 0939-7922 (1969)
16. Ishii H (1978) Spin correlation in dilute magnetic alloys. J Low Temp Phys 32:11
17. Chen K, Jayaprakash C, Krishna-Murthy HR (1987) Spatial correlations around a Kondo impurity. Phys Rev Lett 58:929–932
18. Chen K, Jayaprakash C, Krishnamurthy HR (1992) Spatially dependent zero-frequency response functions and correlation functions in the Kondo model. Phys Rev B 45:5368–5386
19. Sorensen ES, Affleck I (1996) Scaling theory of the Kondo screening cloud. Phys Rev B 53:9153–9167
20. Barzykin V, Affleck I (1996) The Kondo Screening Cloud: What Can We Learn from Perturbation Theory? Phys Rev Lett 76:4959–4962
21. Barzykin V, Affleck I (1998) Screening cloud in the k-channel Kondo model: perturbative and large-k results. Phys Rev B 57:432–448
22. Cornaglia PS, Balseiro CA (2002) Kondo impurities in nanoscopic systems: confinementinduced regimes. Phys Rev B 66:115303
23. Costamagna S, Gazza CJ, Torio ME, Riera JA (2006) Anderson impurity in the one-dimensional Hubbard model for finite-size systems. Phys Rev B 74:195103
24. Molina RA, Weinmann D, Pichard J-L (2006) Interacting electron systems between Fermi leads: effective one-body transmissions and correlation clouds, 0507571. arXiv:condmat/
25. Hand T, Kroha J, Monien H (2006) Spin correlations and finite-size effects in the one-dimensional Kondo Box. arXiv:cond-mat/0602352
26. Borda L (2007) Kondo screening cloud in a one-dimensional wire: numerical renormalization group study. Phys Rev B 75:041307. ISSN: 1098-0121, 1550-235X
27. Sørensen ES, Chang M-S, Laflorencie N, Affleck I (2007) Impurity entanglement entropy and the Kondo screening cloud. J Stat Mech: Theory Exper 2007:L01001–L01001. ISSN: 1742-5468

28. Affleck I, Borda L, Saleur H (2008) Friedel oscillations and the Kondo screening cloud. Phys Rev B 77:180404. ISSN: 1098-0121, 1550-235X
29. Holzner A, McCulloch IP, Schollwöck U, von Delft J, Heidrich-Meisner F (2009) Kondo screening cloud in the single-impurity Anderson model: a density matrix renormalization group study. Phys Rev B 80:205114
30. Büsser CA et al (2010) Numerical analysis of the spatial range of the Kondo effect. Phys Rev B 81:045111
31. Mitchell AK, Becker M, Bulla R (2011) Real-space renormalization group flow in quantum impurity systems: Local moment formation and the Kondo screening cloud. Phys Rev B 84:115120
32. Feng X-Y, Zhang F-C (2011) Kondo spin screening cloud in two-dimensional electron gas with spin-orbit couplings. J Phy: Condens Matter 23:105602. ISSN: 0953-8984, 1361-648X
33. Posske T, Liu C-X, Budich JC, Trauzettel B (2013) Exact results for the Kondo screening cloud of two helical liquids. Phys Rev Lett 110:016602. ISSN: 1079-7114
34. Medvedyeva M, Hoffmann A, Kehrein S (2013) Spatiotemporal buildup of the Kondo screening cloud. Phys Rev B 88:094306
35. Ghosh S, Ribeiro P, Haque M (2014) Real-space structure of the impurity screening cloud in the resonant level model. J Stat Mech: Theory Exper 2014:P04011. ISSN: 1742-5468
36. Florens S, Snyman I (2015) Universal spatial correlations in the anisotropic Kondo screening cloud: analytical insights and numerically exact results from a coherent state expansion. Phys Rev B 92:195106
37. Lee S, Park J, Sim H (2015) Macroscopic quantum entanglement of a Kondo cloud at finite temperature. Phys Rev Lett 114:057203. ISSN: 0031-9007, 1079-7114
38. Nuss M, Ganahl M, Arrigoni E, von der Linden W, Evertz HG (2015) Nonequilibrium spatiotemporal formation of the Kondo screening cloud on a lattice. Phys Rev B 91:085127
39. Ribeiro LC, Martins GB, Gómez-Silva G, Anda EV (2019) Numerical study of the Kondo cloud using finite-U slave bosons. Phys Rev B 99:085139. ISSN: 2469-9950, 2469-9969
40. Diniz G, Diniz GS, Martins GB, Vernek E (2020) Reentrant Kondo effect for a quantum impurity coupled to a metal-semiconductor hybrid contact. Phys Rev B 101:125115. ISSN: 2469-9950, 2469-9969
41. Moca CP, Weymann I, Werner MA, Zaránd G (2021) Kondo cloud in a superconductor. Phys Rev Lett 127:186804. ISSN: 0031-9007, 1079-7114
42. Mukherjee A, Mukherjee A, Vidhyadhiraja NS, Taraphder A, Lal S (2022) Unveiling the Kondo cloud: unitary renormalization-group study of the Kondo model. Phys Rev B 105:085119. ISSN: 2469-9950, 2469-9969
43. Barzykin V, Affleck I (1998) Screening cloud in the k-channel Kondo model: perturbative and large-k results. Phys Rev B 57:432–448
44. Borzenets, I. V. et al. Observation of the Kondo screening cloud. en. Nature, 11 (2020)
45. Affleck I, Simon P (2001) Detecting the Kondo screening cloud around a quantum dot. Phys Rev Lett 86:2854–2857
46. Cornaglia PS, Balseiro CA (2003) Transport through Quantum dots in mesoscopic circuits. Phys Rev Lett 90:216801
47. Pereira RG, Laflorencie N, Affleck I, Halperin BI (2008) Kondo screening cloud and charge staircase in one-dimensional mesoscopic devices. Phys Rev B 77:125327
48. Patton KR, Hafermann H, Brener S, Lichtenstein AI, Katsnelson MI (2009) Probing the Kondo screening cloud via tunneling-current conductance fluctuations. Phys Rev B 80:212403
49. Nishida Y (2013) SU(3) orbital Kondo effect with ultracold atoms. Phys Rev Lett 111:135301
50. Bauer J, Salomon C, Demler E (2013) Realizing a Kondo-correlated state with ultracold atoms. Phys Rev Lett 111:215304
51. Snyman I, Florens S (2015) Robust Josephson-Kondo screening cloud in circuit quantum electrodynamics. Phys Rev B 92:085131
52. Affleck I (2009) The Kondo screening cloud: what it is and how to observe it. arXiv
53. Park J, Lee S, Oreg Y, Sim H (2013) How to directly measure a kondo cloud's length. Phys Rev Lett 110:246603. ISSN: 0031-9007, 1079-7114

54. Moca CP, Weymann I, Werner MA, Zaránd G (2021) Kondo cloud in a superconductor. Phys Rev Lett 127:186804
55. Erpenbeck A, Cohen G (2021) Resolving the nonequilibrium Kondo singlet in energyand position-space using quantum measurements. SciPost Phys 10
56. Figgins J et al (2019) Quantum engineered Kondo lattices. Nat Commun 10:5588. ISSN: 2041-1723
57. Prüser H et al (2011) Long-range Kondo signature of a single magnetic impurity. Nat Phys 7:203–206. ISSN: 1745-2473, 1745-2481
58. Jiang Y, Zhang YN, Cao JX, Wu RQ, Ho W (2011) Real-space imaging of Kondo screening in a two-Dimensional O2 lattice. Science 333:324–328
59. Seaman CL et al (1991) Evidence for non-Fermi-liquid behavior in the Kondo alloy Y1-xUxPdd. Phys Rev Lett 67:2882–2885
60. MacLaughlin DE et al (2004) Disorder, inhomogeneity and spin dynamics in f-electron non-Fermi liquid systems. J Phys: Condens Matter 16:S4479–S4498
61. Aronson MC et al (1995) Non-Fermi-liquid scaling of the magnetic response in UCu5-xPdx(x = 1, 1.5). Phys Rev Lett 75:725–728
62. Bernal OO, MacLaughlin DE, Lukefahr HG, Andraka B (1995) Copper NMR and thermodynamics of UCu5-xPdx: evidence for Kondo disorder. Phys Rev Lett 75:2023–2026
63. Löhneysen Hv et al (1994) Non-Fermi-liquid behavior in a heavy-fermion alloy at a magnetic instability. Phys Rev Lett 72:3262–3265
64. Maple MB et al (1994) Non-Fermi liquid ground states in strongly correlated f-electron materials. J Appl Phys 76:6137–6137
65. Dobrosavljević V, Kirkpatrick TR, Kotliar BG (1992) Kondo effect in disordered systems. Phys Rev Lett 69:1113–1116
66. Miranda E, Dobrosavljevic V, Kotliar G (1997) Disorder-driven non-Fermi liquid behavior in Kondo alloys. Phys Rev Lett 78:290–293. ISSN: 0031-9007, 1079-7114
67. Cornaglia PS, Grempel DR, Balseiro CA (2006) Universal Distribution of Kondo Temperatures in Dirty Metals. Phys Rev Lett 96:117209
68. Kettemann, S. & Mucciolo, E. R. Free magnetic moments in disordered metals. en. Journal of Experimental and Theoretical Physics Letters 83, 240-245. issn: 0021- 3640, 1090-6487 (2006)
69. Kaul RK, Ullmo D, Baranger HU (2003) Mesoscopic fluctuations in quantum dots in the Kondo regime. Phys Rev B 68:161305. ISSN: 0163-1829, 1095-3795
70. Nagaoka Y (1965) Self-consistent treatment of Kondo's effect in dilute alloys. Phys Rev 138:A1112–A1120
71. Milovanović M, Sachdev S, Bhatt RN (1989) Effective-field theory of local-moment formation in disordered metals. Phys Rev Lett 63:82–85. ISSN: 0031- 9007
72. Martin I, Wan Y, Phillips P (1997) Size dependence in the disordered Kondo problem. Phys Rev Lett 78:114–117. ISSN: 0031-9007, 1079-7114
73. Miranda E, Dobrosavljević V (2001) Localization-induced Griffiths phase of disordered Anderson lattices. Phys Rev Lett 86:264–267. ISSN: 0031-9007, 1079-7114
74. Cornaglia PS, Balseiro CA (2002) Kondo impurities in nanoscopic systems: confinementinduced regimes. Phys Rev B 66:115303
75. Micklitz T, Altland A, Costi TA, Rosch A (2006) Universal dephasing rate due to diluted Kondo impurities. Phys Rev Lett 96:226601. ISSN: 0031-9007, 1079-7114
76. Kettemann S, Mucciolo ER, Varga I, Slevin K (2012) Kondo-Anderson transitions. Phys Rev B 85:115112. ISSN: 1098-0121, 1550-235X
77. Kettemann S, Mucciolo ER (2007) Disorder-quenched Kondo effect in mesoscopic electronic systems. Phys Rev B 75:184407. ISSN: 1098-0121, 1550-235X
78. Zhuravlev A et al (2007) Nonperturbative scaling theory of free magnetic moment phases in disordered metals. Phys Rev Lett 99:247202. ISSN: 0031- 9007, 1079-7114
79. Bergmann G, Thompson RS (2014) The range of the Kondo cloud in weakly disordered hosts. arXiv:1410.7512 [cond-mat]

80. Miranda VG (2014) Dias da Silva LGGV, Lewenkopf CH (2014) Disorder-mediated Kondo effect in graphene. Phys Rev B 90:201101. ISSN: 1098-0121, 1550- 235X
81. Moure N, Lee H-Y, Haas S, Bhatt RN, Kettemann S (2018) Disordered quantum spin chains with long-range antiferromagnetic interactions. Phys Rev B 97:014206. ISSN: 2469-9950, 2469-9969
82. Park K-Y, Jang I, Kim K-S, Kettemann S (2021) Inhomogeneous Kondo destruction by RKKY correlations. Ann Phys 435:168501. ISSN: 00034916
83. Sen S, Vidhyadhiraja NS, Jarrell M (2018) Emergence of non-Fermi liquid dynamics through nonlocal correlations in an interacting disordered system. Phys Rev B 98:075112
84. Slevin K, Kettemann S, Ohtsuki T (2019) Multifractality and the distribution of the Kondo temperature at the Anderson transition. Eur Phys J B 92:281. ISSN: 1434-6028, 1434-6036
85. Beenakker CWJ (1997) Random-matrix theory of quantum transport. Rev Mod Phys 69:731–808
86. Liu DE, Burdin S, Baranger HU, Ullmo D (2012) Mesoscopic Anderson box: connecting weak to strong coupling. Phys Rev B 85:155455

Part II
Quantum Impurity Problems and Natural Orbitals

Chapter 3
Few-Body Nature of Kondo Correlated Ground States

3.1 Motivations

In the previous part, it was discussed that despite the apparent simplicity of quantum impurity models, they hide non trivial physics and their solution is in general a hard task. Indeed, the many-body description of these problems forces one to work in the exponentially large Fock space where every state, or Slater determinant for fermionic systems that we consider, is treated equally. It could be hoped that most of this huge amount of determinants do not participate notably to the ground state characterization and it would be appealing not to consider them. For this purpose, it is desirable to find the corner of the Fock space within which only the most relevant determinants remain in the computation, changing the exponential scaling to some polynomial function of the system size.

Quantum impurity problems, due to the sparsity of the interaction between Wannier (real space) orbitals, disclose simplicity when they are expressed with *natural orbitals* (NOs) through a given unitary transformation, that is defined in this chapter. Natural orbitals are widely used in the quantum chemistry community [1–5] to quantify electronic correlations between molecular spin-orbitals. Although there is no obvious resemblance between molecules and impurity problems, they both contain a hierarchical structure of correlations that is revealed by natural orbitals. Bridges have been built between these two fields thanks to DMFT, that was investigated to describe strongly correlated materials beyond the usual approximations of quantum chemistry (density functional theory, local-density approximation, etc.) [6–8]. To efficiently simulate the underlying impurity problems of DMFT, exact diagonalization (or full configuration interaction in quantum chemistry) can be dressed up with various approximations, and NOs-based approaches originally suited for molecules have thus been democratized for impurity problems [9–19].

As Kondo-correlations become stronger, the scattering from electronic reservoirs generates the long ranged Kondo cloud presented in Chap. 2, and we expect particles participating to the screening process to be correlated with each other. However, the question concerning the number of correlated particles (that *dominantly* participate to

M. Debertolis, *Quantum Impurity Problems in the Framework of Natural Orbitals*, Springer Theses, https://doi.org/10.1007/978-3-031-47233-6_3

correlations) in this state has not been addressed extensively, and natural orbitals are a powerful tool to answer it. Formally, NOs are eigenvectors of the one-body density matrix Q, and their corresponding eigenvalues λ_n (that can take values between 0 and 1 for spinless fermions, corresponding to the NO occupancy) quantify the correlation of each orbital in the given state. For the purpose of this study, we will consider the ground state of several quantum impurity models. In these problems, eigenvalues of the one-body density matrix are exponentially close to the extreme values 0 or 1, such that the corresponding orbital is almost empty or full.

A mathematical study concerning the complexity of quantum impurities [20] proved this exponential behavior, and introduced an upper bound for the exponential decay of $Q(1-Q)$ (that folds the part of the eigenvalues close to 1 near 0):

$$\sigma_n \leq c \exp\left[-\frac{n}{14m\log(2\omega^{-1})}\right], \tag{3.1}$$

where c is a universal constant, ω the spectral gap of the bath and m the size (i.e. the number of degrees of freedom) of the impurity. This exponential decay supposes that only a small number of orbitals are relevant in the ground state, that becomes less complicated to deal with. Recent numerical studies proposed a stronger statement, that only a single orbital would be necessary to describe the Kondo state in the strong and weak coupling limits [13, 15], such that the full many-problem would become almost trivial. However, it will be shown in this chapter that the problem is not that simple and that results reported in these studies are not valid in the thermodynamic limit. An extensive NRG study is performed to characterize correlations in different regimes, especially where the Kondo temperature goes to zero, when correlations fully develop through the entire reservoir.

The results presented in this chapter are published in Ref. [17] and the structure will follow that of the article. First, generalities concerning the one-body density matrix and natural orbitals are discussed. Then, the spectrum of this matrix is investigated in the Kondo regime of the IRLM and the SIAM via NRG simulations. Finally, an ansatz for the ground state wave function based on natural orbitals is presented, and its precision is discussed as the interaction strength changes.

3.2 One-Body Density Matrix

The one-body density matrix is defined in any basis as the collection of one-body correlators involving each possible spin-orbital. For spinless fermions of the IRLM, this matrix is defined as:

$$Q_{ij} = \left\langle c_i^\dagger c_j \right\rangle. \tag{3.2}$$

In the literature, it is sometimes called the covariance matrix, or the correlation matrix. Operators can be evaluated in any state, yet the ground state will be the main interest in this work. Fermion operators can be defined for any basis, such that matrix elements of Q depend on the starting orbital set. Here, the basis of Wannier orbitals is considered, and since the impurity is also taken into account here, the labelling starts at -1 with $c_{-1} = d$. Diagonalizing this matrix in any basis leads to the same set of eigenvectors, the so-called natural orbitals. The diagonalization is performed by the application of the unitary matrix D, such that in that case natural orbitals q_n are linear combinations of the physical orbital c_i:

$$q_n = \sum_{i=-1}^{N-1} D_{in} c_i \,, \quad \text{with} \quad \sum_{k=0}^{N} Q_{ik} D_{kn} = \lambda_n D_{in}. \tag{3.3}$$

Eigenvalues of Q have been introduced as λ_n, and they represent the occupancy of the nth natural orbital in the ground state:

$$\lambda_n = \langle q_n^\dagger q_n \rangle = \sum_{i,j} D_{in} D_{jn} \langle c_i^\dagger c_j \rangle. \tag{3.4}$$

From this definition, it is clear that $0 < \lambda_n < 1$ for the spinless fermions of the IRLM. If $\lambda_n = 0$, the orbital q_n does not appear in the ground state at all, while $\lambda_n = 1$ means that it is present in every determinant constituting the ground state and therefore trivially factorizes (full orbital). Two extreme cases are studied now to gain intuition about the spectrum of the matrix Q: the case of free electrons described by a single Slater determinant, and the simplest case of entangled particles represented by a Bell-like state.

3.2.1 *Single Slater Determinant*

Consider here a translation-invariant tight-binding Hamiltonian of free fermions:

$$H_{\text{free}} = \sum_{\langle\langle i,j \rangle\rangle} t \left(c_i^\dagger c_j + \text{h.c.} \right). \tag{3.5}$$

$\langle\langle i, j \rangle\rangle$ is used for nearest neighbor hopping in this example, but can be more general. This Hamiltonian is diagonalized by plane waves, defined as the Fourier transform of real-space orbitals: $c_k = \sum_j e^{ikj} c_i$, and their dispersion relation is written ϵ_k. In this case, at $T = 0$, the ground state is a single Slater determinant that reads:

$$|\psi_0\rangle = \prod_{\epsilon_k \leq 0} c_k^\dagger |0\rangle. \tag{3.6}$$

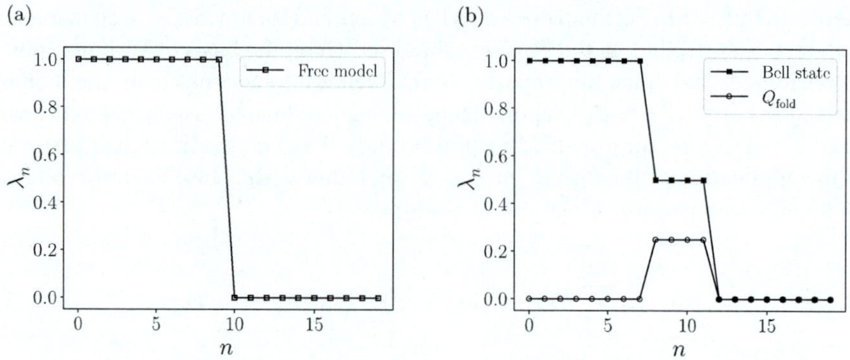

Fig. 3.1 **a** Spectrum of Q for a free model of 20 sites, sorted in descending order. The Fermi energy is in the middle of the spectrum, such that there are as many filled orbitals as empty ones. The straight line is a guide for the eye. **b** Spectrum of Q in a Bell-like state involving four different orbitals above a Fermi sea of 16 energy modes. The spectrum of the folded matrix $Q_{\text{fold}} = Q(1 - Q)$ is also plotted

In this basis, it is plain that $\langle\psi_0|c_k^\dagger c_{k'}|\psi_0\rangle = \delta_{kk'}\Theta\left(-\epsilon_k\right) = n_k\delta_{kk'}$, where Θ is the Heaviside step function with the convention $\Theta(0) = 1$. The Fermi energy is chosen to be $E_{\text{F}} = 0$ without loss of generality. Hence, it is straightforward to see that plane waves diagonalize the correlation matrix of this problem, with the spectrum $\lambda_k = n_k$, which is depicted in the left panel of Fig. 3.1.

3.2.2 Simple Structure of Entanglement: Bell-Like State

Consider now a more complicated state, which is a sum of two Slater determinants. This is a kind of fermionic Bell state, that usually describes the entanglement between two qubits. Such a state reads:

$$|\psi_{\text{Bell}}\rangle = \frac{1}{\sqrt{2}}\left(c_a^\dagger c_b^\dagger + c_c^\dagger c_d^\dagger\right)|\phi\rangle. \tag{3.7}$$

$|\phi\rangle$ is a Slater determinant that does not involve orbitals a, b, c, and d and acts as a background for the entangled state. These orbitals appear in distinct pairs in the wave function, so that their occupancy is neither 0 nor 1 as in the previous case. The Q matrix is nevertheless diagonal in this basis, but the occupancies of these 4 orbitals are here $1/2$, which reflects the strong entanglement between them. The spectrum of Q for this state is plotted in the right panel of Fig. 3.1. Eigenvalues are sorted in descending order, and values of λ_n at 0 and 1 strictly correspond to filled or empty orbitals in $|\phi\rangle$. It is useful to introduce the folded Q matrix here, which provides a sensitive measure of correlation between particles:

$$Q_{\text{fold}} = Q(1 - Q). \tag{3.8}$$

Indeed, eigenvalues of Q_{fold} far from 0 corresponds to correlated orbitals, and their value $\tilde{\lambda}_n$ corresponds to a degree of entanglement, between 0 for a filled or empty orbital bearing no correlations and 0.25 for a highly entangled Bell-like state. In the following, the spectrum of Q_{fold} will be used for readability. Note that its eigenvectors are still natural orbitals, but its eigenvalues are no more occupancies of the latter.

3.2.3 Some General Properties of the Q Matrix

In the IRLM, the correlation matrix has to be calculated in the many-body ground state that is *a priori* unknown. For this purpose, an NRG simulation of this model (and later on of the SIAM) is developed to compute Q. It is worth pointing out that most NRG studies of quantum impurities focus on observables defined with the impurity degrees of freedom. The matrix Q involves rather observables that are hybridized between the environment and the impurity degrees of freedom, and implementing such observables is more demanding. At step N, there are N^2 operators that follow the RG flow, which renders computations quite expensive as the chain grows. However, for the IRLM and the SIAM, the many-body states are constructed with a reasonable computational cost, and this makes this study feasible. The details of the RG scheme for the IRLM and the SIAM, and the implementation of their respective one-body density matrices are given in Appendices A and B.

The IRLM Hamiltonian manifests particle-hole symmetry at $\epsilon_d = 0$, and is invariant under the application of the particle-hole conjugation operator:

$$P = \prod_{i=0}^{\frac{N}{2}-1} \left(c_{2i-1} - c_{2i-1}^{\dagger}\right)\left(c_{2i} + c_{2i}^{\dagger}\right), \tag{3.9}$$

such that $P^{\dagger}\mathcal{H}_{\text{IRLM}}P = \mathcal{H}_{\text{IRLM}}$. This operator transforms annihilation (creation) operators in their complex conjugate as $P^{\dagger}c_i P = (-1)^i c_i^{\dagger}$. Its action on a state is the same, every electron becomes a hole, and in particular $P^{\dagger}|0\rangle = c_{-1}^{\dagger}...c_{N-1}^{\dagger}|0\rangle$. Its unitarity imposes $P^2 = 1$, and its eigenvalues are thus ± 1. Hence, applying P to every operator in Q leads to the following property:

$$Q_{ij} = \delta_{ij} - (-1)^{i+j} Q_{ji}. \tag{3.10}$$

However, $\mathcal{H}_{\text{IRLM}}$ is hermitian and has real eigenvectors, such that Q should be symmetric, $Q_{ij} = Q_{ji}$. To satisfy both constraints, entries of Q form a chessboard, with $Q_{ij} = 0$ for even values of $i + j$, with $i \neq j$. PH symmetry also imposes that every diagonal element Q_{ii} is $1/2$: in the ground state, each Wannier orbital is half-filled. It also implies that the spectrum of Q is symmetric around $1/2$, such that natural orbitals come in pairs with $\lambda_n = 1 - \lambda_{N-n}$. This provides a useful check of the particle-hole symmetry within numerical simulations, that can be slightly broken by numerical noise if it is not strictly imposed. As one wants to explore the PH symmetric and the symmetry broken phases, PH symmetry is not imposed in the simulations and only the conservation of the number of particles is used. In the simulation of the SIAM, conservation of the total spin is also implemented.

3.3 IRLM and SIAM Correlation Spectra

The properties of the correlation spectra of both the IRLM and SIAM are discussed here. These two models are expected to share the same universal low energy physics, in the charge sector of the IRLM and in the spin one of the SIAM. The study of the IRLM, however, may not be universal enough since its fermions are related to the original degrees of freedom in a complicated fashion, and the study of the SIAM is performed to enforce generality of the results for ground states of quantum impurity problems. Using the Wilson chain discretization required by NRG simulations, expressions of these Hamiltonians are recalled, here at PH symmetry:

$$
\begin{aligned}
\mathcal{H}_{\text{IRLM}} &= U\left(d^\dagger d - \frac{1}{2}\right)\left(c_0^\dagger c_0 - \frac{1}{2}\right) + V\,(d^\dagger c_0 + c_0^\dagger d) + \sum_{i=0}^{N-1} t_i\,\left(c_{i+1}^\dagger c_i + c_i^\dagger c_{i=1}\right),\\
\mathcal{H}_{\text{SIAM}} &= U\left(d_\uparrow^\dagger d_\uparrow - \frac{1}{2}\right)\left(d_\downarrow^\dagger d_\downarrow - \frac{1}{2}\right) + V\sum_{\sigma=\uparrow,\downarrow}^{V} (d_\sigma^\dagger c_{0\sigma} + c_{0\sigma}^\dagger d_\sigma)\\
&\quad + \sum_{\sigma=\uparrow,\downarrow}\sum_{i=0}^{N-1} t_i\,\left(c_{i+1\sigma}^\dagger c_{i\sigma} + c_{i\sigma}^\dagger c_{i+1\sigma}\right).
\end{aligned}
\tag{3.11}
$$

Since an exponential decay of eigenvalues of the one-body density matrix of both models is expected, elements of Q are calculated at a high precision to obtain the best resolution for the largest number of eigenvalues. For this purpose, an exhaustive study concerning the convergence of eigenvalues, in addition to the convergence to the thermodynamic limit of both simulations is presented in the next Sect. 3.4.

The ground state of the SIAM is a spin singlet, so that the different spin sectors do not mix and $\langle c_{i\sigma}^\dagger c_{j\sigma'}\rangle \propto \delta_{\sigma\sigma'}$, and $\langle c_{i\uparrow}^\dagger c_{j\uparrow}\rangle = \langle c_{i\downarrow}^\dagger c_{j\downarrow}\rangle$. Thus, only the calculation of $\langle c_{i\uparrow}^\dagger c_{j\uparrow}\rangle$ is required, and an additional two-fold degeneracy is expected in the SIAM with respect to the IRLM due to this spin symmetry. Another convention for the Q matrix of the SIAM, would have been to include spin within orbitals such that each entry of Q would have taken values between 0 and 2, and the two fold degeneracy would have disappeared. This definition is not used to maintain similarities with the IRLM results, and it is only a matter of labels.

The spectra are displayed in the top left panel of Fig. 3.2 for the IRLM and in the top right panel for the SIAM. NRG simulations for the IRLM use $\Lambda = 1.5$ for the discretization parameter, a tunneling $V = 0.15$ leading to a hybridization $\Gamma \simeq 3 \times 10^{-2}$ in units of half-bandwidth $D = 1$, and $N = 180$ sites on the Wilson chain. The lowest eigenenergies of the Hamiltonian are thus of the order $1.5^{-90} \simeq 10^{-16}$, that is well below any Kondo scale considered in this study. Parameters for the SIAM are $\Lambda = 1.5$, $\Gamma = 10^{-2}$ and $N = 180$, with twice the number of orbitals because of the spin. The figure plots $1 - \lambda_n$ on the left part of the spectra $1/2 < \lambda_n < 1$, and λ_n on the right part $0 < \lambda_n < 1/2$. It is similar to plot the spectrum of Q_{fold}, but the eigenvalues are true occupancies here. The abscissa axis has been relabeled to be PH symmetry-friendly: the left part ranges from $-N/2$ to -1, and the opposite for the right part such that orbital 1 is the PH conjugate of the -1 one. The vertical axis is given in log-scale to reveal the exponential decay.

For each value of the interaction considered in the top left panel of Fig. 3.2, the spectrum exhibits the expected exponential decay [20], on top of an approximate four-fold degeneracy of the four largest eigenvalues. These four highly-correlated states indicate a Bell-like entanglement between the concerned natural orbitals q_{-2}, q_{-1}, q_1 and q_2. This suggests that at negative U, the impurity and the first energy shell are either both filled or empty due to Coulomb interaction, and at positive U one is full and the other empty or vice-versa. Every other orbital has its occupancy departing exponentially from 1/2, and by writing in a simpler way Eq. (3.1), this exponential decay can be parametrized by a single parameter x as:

$$\begin{cases} \lambda_n \propto e^{-xn} & \text{for} \quad n > 2, \\ 1 - \lambda_n \propto e^{+xn} & \text{for} \quad n < -2. \end{cases} \tag{3.12}$$

This parameter x corresponds to the decay rate of eigenvalues, and varies with the amount of correlations in the system. We can examine how it depends on the Kondo temperature T_{K}: the bottom left panel of Fig. 3.2 displays this decay rate as a function of T_{K}, that is vanishing for $U \to U_c \simeq -\pi(\sqrt{2} - 1)$ in the IRLM: as T_{K} is quenched to zero, x becomes smaller, corresponding to a slower decay of eigenvalues and hence a more correlated state.

The same behavior is reported for the SIAM on the top right and bottom right panels of Fig. 3.2. However, in this model the spin degeneracy induces a decay rate x twice smaller compared to the IRLM for the same T_{K} (that is also computed with the susceptibility on the impurity). The four highly entangled orbitals are also present in the ground state, which also indicates a Bell-like state to appear around the impurity, but these spectra are not enough to describe it precisely. Its structure will be investigated in the next sections, and it will be shown that the simple picture of a single orbital coupling to the impurity breaks down in the thermodynamic limit.

The behavior of the largest eigenvalue λ_{max} (in the range [0, 1/2]), and of the decay rate x are reported in Fig. 3.3 over the full range of the IRLM interaction U. λ_{max} remains small when $U > 0$ (the inset shows that it increases slightly up to the half bandwidth $D = 1$, and decreases again after), and cancels at $U = 0$, when the

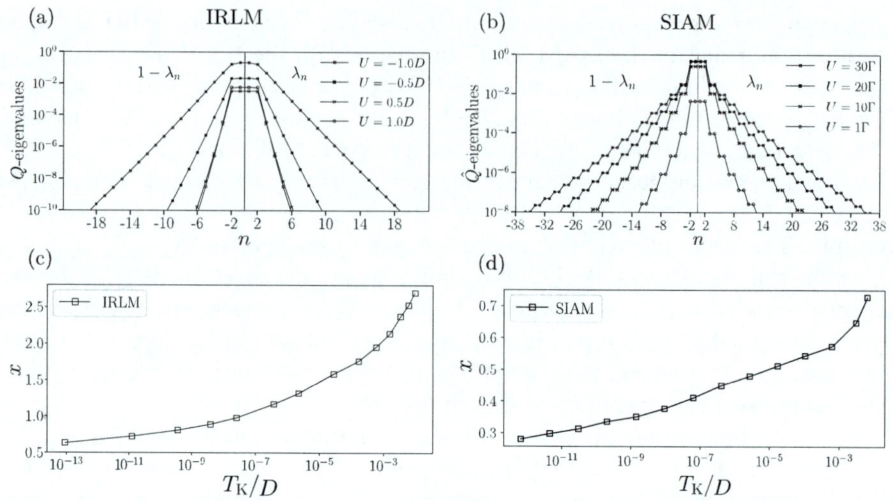

Fig. 3.2 **a** Spectrum of Q for the IRLM at different values of the interaction U in both regimes $U > 0$ and $U < 0$, for $V = 0.15D$ (that implies an hybridization $\Gamma = 10^{-3}D$). The left part of the plot shows $1 - \lambda_n$ for $1/2 < \lambda_n < 1$, to exhibit clearly PH symmetry. The x-axis has been relabeled, such that opposite n corresponds to PH conjugate orbitals. **b** Same plot as **a** for the correlation spectrum of the SIAM, with $\Gamma = 10^{-2}D$. **c** Decay rate x of Q eigenvalues against the Kondo temperature, defined in the correlated regime of the IRLM $U_c < U < 0$. **d** Decay rate x of Q eigenvalues against the Kondo temperature for the SIAM, defined for $U > 0$. The spin degeneracy gives a decay rate x that is roughly halved compared at the one in the IRLM. Another definition of Q including the spin in each orbital would have given the same x, but occupancies would have taken values between 0 and 2

state is trivial. It is the case since the impurity is accounted in the matrix Q: if the d-level would have been separated from the bath in Q, the state at $U = 0$ with natural orbitals would be a sum of determinants, leading to the same state but in a more complex fashion, that is avoided in our definition of Q. In the correlated regime of the IRLM $U_c < U < 0$, $\lambda_{\max} \to 0.5$, approaching its maximal value. After the quantum phase transition at $U = U_c$, the maximal Q eigenvalue discontinuously falls to small values, indicating a weakly correlated regime.

The decay rate x reports the same behavior in the parameter space. An important observation here lies at the transition: as the bottom left panel of Fig. 3.2, x seemingly vanishes at the transition. Hence, in the quantum critical regime, where fluctuations are maximal, the state is truly many-body and every state carries the same amount of correlations. However, even for an exponentially small T_K, x is finite and only *few* natural orbitals (few is defined more quantitatively in the next section) are relevant for correlations, while the contribution of the other orbitals are orders of magnitude below.

A common probe for Kondo correlations is the perturbation of the system at the Kondo scale, allowing to investigate the response of the system. In the IRLM, a

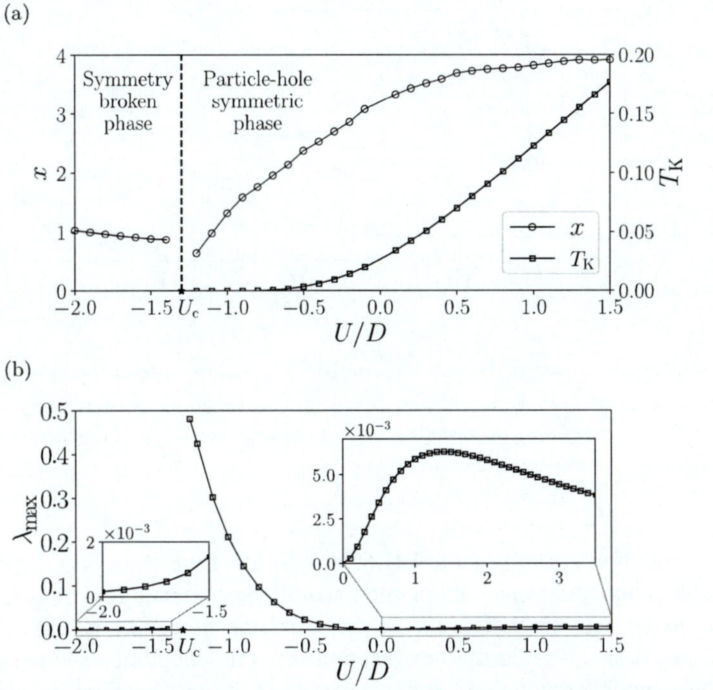

Fig. 3.3 **a** Decay rate of Q eigenvalues and of the Kondo temperature of the IRLM as a function of U. For both **a** and **b** plots $V = 0.15$, and the $U = 0$ point corresponding to the free resonant level model is not plotted (the curve is nevertheless continuous near this point). **b** Largest eigenvalue λ_{max} of the IRLM matrix Q as a function of interaction U. Insets zooms on parts where $\lambda_{\text{max}} \simeq 0$ to see the structure in the weakly-correlated regime $U > 0$. U_c corresponds to the transition point to the PH symmetry broken phase, that is the antiferromagnetic-ferromagnetic transition in the Kondo model

biasing potential $\epsilon_d d^\dagger d$ (corresponding to a Zeeman splitting between up and down spins in the Kondo model) of order T_{K} is applied to prevent the formation of the singlet. The occupancy $n_d = \langle d^\dagger d \rangle$ is very sensitive to symmetry breaking (leaving its $1/2$ value), and the extreme cases $n_d = 0$ (for $\epsilon_d \gg T_{\text{K}}$) and $n_d = 1$ (for $\epsilon_d \ll T_{\text{K}}$) remove any kind of dynamics of the impurity, thus killing all correlations. Therefore, breaking PH symmetry reduces correlations in the system, and the decay rate x will be used to measure this reduction. In Fig. 3.4, both n_d and x are plotted against the perturbation ϵ_d, that takes values below and above T_{K}: $\epsilon_d / T_{\text{K}} \in [10^{-2}, 10^2]$. When $\epsilon_d \ll T_{\text{K}}$, the perturbation is not felt by the system and the slight symmetry breaking does not affect correlations. As soon as the Kondo scale is reached (and then exceeded), the perturbation is strong enough to suppress correlations, and x increases while the impurity gets polarized ($n_d \to 0$ or 1, depending on the sign of ϵ_d).

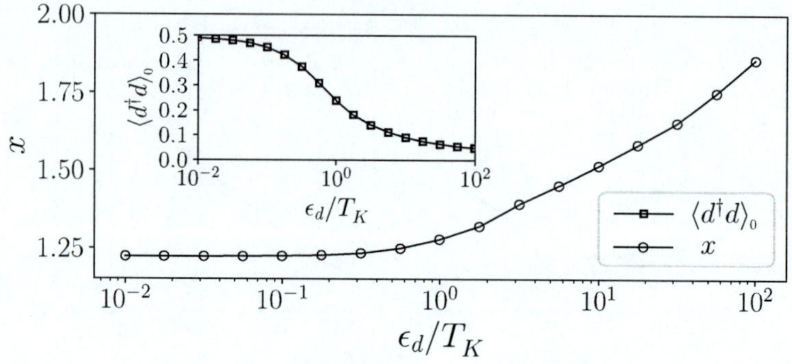

Fig. 3.4 Decay rate x versus the on-site perturbation ϵ_d on the d-level rescaled by the Kondo temperature, at $U = -1.0$, $V = 0.15$ and $\Lambda = 1.5$. The Kondo temperature at $\epsilon_d = 0$ is $T_K = 1.87 \times 10^{-6}$. The inset shows the occupancy of the impurity level n_d as a function of the rescaled ϵ_d, and displays a typical Kondo crossover

The correlation spectra of the IRLM and SIAM have been investigated in this section, but some questions remain open. Could the observed few-body behavior of correlations be only an artifact of the discretization procedure used for numerical simulations. It is shown in the next section that this potential issue is being ruled out from numerical details on the convergence of the results. Finally, an ansatz for the wave function using the few-body structure of correlations will be proposed, and will be confronted to the simplistic picture of a single orbital coupled to the impurity put forward by previous studies [15, 21].

3.4 Convergence to the Thermodynamic Limit

The thermodynamic limit of any NRG simulation on the Wilson chain is strictly reached when $N \to \infty$ and $\Lambda \to 1$. In the IRLM, the one-dimensional continuous conduction band with a box density of states is recovered, as well as the true hybridization of the impurity to a continuum. However, actual simulations are performed at finite N and at $\Lambda > 1$, that should be analyzed properly to reach the thermodynamic limit with accurate enough results for a surmountable numerical cost.

The numerical cost can be reduced by another parameter of the simulation, which is not explicitly part of the Hamiltonian: the number of kept states in Fock space sectors N_{kept}. Tuning this parameter is also important, since it also affects accuracy of the constructed state and thus of every observable. The convergence with respect to these three parameters is investigated here.

In Fig. 3.5, the convergence against chain length N for fixed Λ is shown. The top panel shows the Q spectrum (for eigenvalues in the range $0 < \lambda_n < 1/2$) of the IRLM for different system sizes N. When the chain is long enough, eigenvalues

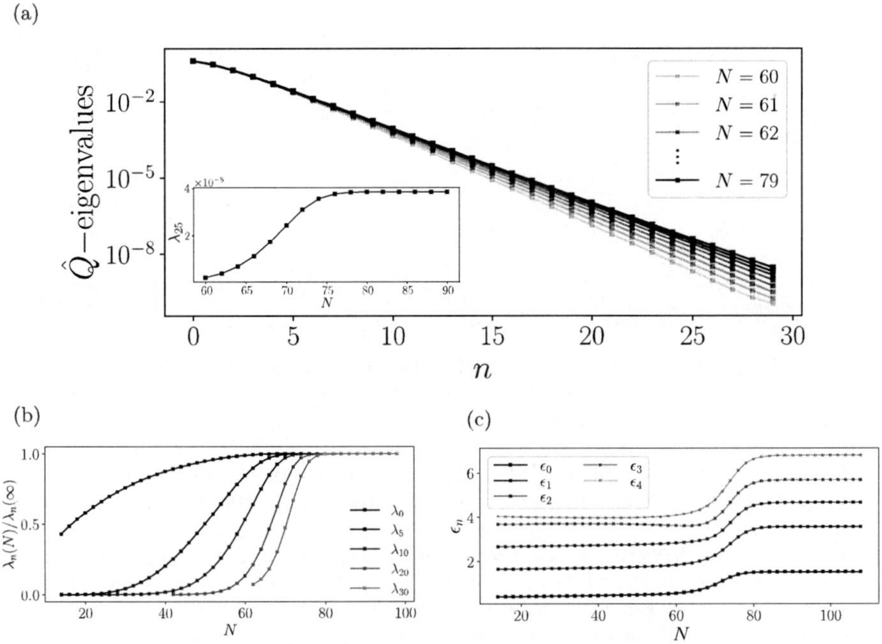

Fig. 3.5 **a** IRLM eigenvalues of Q for $U = -1.2$ and $\Lambda = 2.25$ for various lengths N of the Wilson chain. The inset shows the convergence of the 25th eigenvalue (starting from the middle of the spectrum as usual). **b** Same plot as the inset of **a**, with the same data but for different eigenvalues appearing during the flow. **c** Flow of the five lowest eigenenergies of the rescaled Hamiltonian. There is a clear crossover (similar to that of λ_N) to the Kondo fixed point at scales around $N = 75$, leading to an approximate Kondo temperature of $T_K \simeq 2.25^{-75/2} \simeq 10^{-13}$ as expected

appear to be converged to the infinite chain limit. The inset and the bottom left panel show the evolution of some eigenvalues during the flow. When a new natural orbital enters in the system (at step N for eigenvalue $n = N/2$ in the IRLM, or eigenvalue $n = N$ in the SIAM) its occupancy λ_n changes until it reaches a fixed value. Before reaching system sizes leading to energies of the order of the Kondo temperature T_K, this evolution is relatively slow, and quickens when this energy scale is reached. The bottom right panel shows the evolution of the first five eigenvalues of the Hamiltonian evolving along the flow. The crossover energy from the short range value to the fixed point corresponds to the Kondo temperature, which in this case ($U = -1.2$, $\Lambda = 2.25$) is around $\Lambda^{-75/2} \simeq 10^{-13}$. This extreme regime is investigated here to show that such correlated states can be converged, to ensure that every set of parameters of the study (for are less extreme values of T_K) are converged. Note that it is not possible to obtain eigenvalues $\lambda_{n>N/2}$, that are not defined in the flow yet and that would require to consider longer chain. For the purpose of the study, only the first tens of orbitals are required, such that N is chosen only for the thermodynamic limit to be reached. For the convergence with respect to N and Λ only the results of the

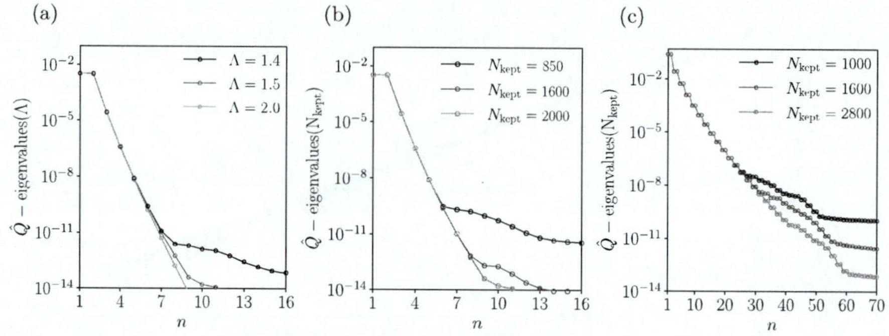

Fig. 3.6 **a** Spectrum of the IRLM Q matrix for different discretizations of the bath monitored by Λ, for $U = 0.5$ and a total number of states kept $N_{kept} = 2000$. As Λ is taken closer to 1, thus approaching the continuous limit, small eigenvalues are less converged for fixed $N_{kept} = 2000$. This is an artifact coming from the collapse of energy shells, which is solved by keeping more states at each step. **b** Same plot as **a**, but at fixed $\Lambda = 1.5$ for various number of kept states. The issues reported in **a** are seen to be solved as N_{kept} increases. **c** Spectrum of the correlation matrix of the SIAM against the number of kept states, for $\lambda = 2.0$ and $U = 20\Gamma$. Only this study will be shown to compare to the IRLM in order not to overburden the text, but every other convergence test has been done in this model and the figures match. In the SIAM, the number of states required to reach the same accuracy is much larger, demonstrating the practical advantage of using the IRLM to describe low energy properties of impurity models, when possible

IRLM are shown as they are similar for the SIAM, and every figure is not plotted twice for obvious reasons of convenience.

The convergence with respect to the discretization parameter Λ is shown in the left panel of Fig. 3.6 for the IRLM. The first few eigenvalues of Q are shown for $\Lambda = 1.4$, 1.5 and 2.0, at fixed N_{kept} and N. For Λ being too close to one, it seemingly appears that the exponential decay does not survive in the tails of the spectrum which level off. Actually, when Λ approaches 1, the energy difference between two consecutive t_n is so small, that every shell (corresponding to a step of the flow) is too close in energy to the next one. This brings convergence issues when the number of kept states is not sufficiently large: high energy states are thrown away, but they were not sufficiently decoupled from the low energy states of further iterations, such that one accumulates large errors. To compensate for this collapse of energy shells, it is thus necessary to keep more states along the flow, burdening the simulations.

The convergence with N_{kept} is shown in the center panel of Fig. 3.6 for fixed Λ and N. The deviation from the exponential decay is only seen when N_{kept} is too small, and the spectrum is recovered in its entirety when enough states are kept. For a better understanding, the total number of kept states is given instead of the number of kept states in each sector. The truncation procedure is explained in Appendices A and B for the IRLM and the SIAM respectively. The convergence against N_{kept} in the SIAM is plotted in the right panel of Fig. 3.6: it appears that this model is harder to converge, and the larger number of sectors (each sector of fixed number of particles also have sub-sectors of fixed spin density) requires to keep a larger total number of states in order to converge the ground state at the same accuracy. Nevertheless, exponential decay seems robust.

Thus, for the simulation performed in this study, $\Lambda = 2.0$, $N_{\text{kept}}^{\text{tot}} \simeq 400$ and $N = 110$ sites are good parameters to reach the thermodynamic limit with high accuracy for the spinless model, for eigenvalues of Q above the machine precision $\simeq 10^{-16}$. It is important to stress that simulations for $U \gtrsim U_c$ have been individually checked and are fully converged.

3.5 Few-Body Ansatz from Natural Orbitals

The correlation spectrum of impurity models has been shown to be highly hierarchical in almost all the parameter space, except for T_{K} being too close to zero. However, the exponential decay stays relatively fast in realistic Kondo regimes, where $T_{\text{K}}/D \simeq 10^{-4} - 10^{-2}$. With these results, it is established that ground states of quantum impurities are few-body in nature for practical reasons. Indeed, almost all the eigenvalues λ_n of Q are exponentially close to 0 or 1, so that a *good* approximation is to assume that their associated natural orbital is totally uncorrelated. In other words, a certain number of orbitals whose folded eigenvalues $\lambda_n(1 - \lambda_n)$ lie below an arbitrary threshold are frozen in the ground state, keeping a core of M *active*, or *correlated* orbitals within the ground state wave function. These orbitals have occupancies the closest to $1/2$, and their number M is chosen to be even in order to respect PH symmetry, so that the correlated sector remains exactly half-filled. The $N - M$ frozen orbitals are described by a single Slater determinant $|\Psi_0\rangle = \prod_{m=-\frac{N}{2}}^{-\frac{M}{2}-1} q_m^\dagger \, |0\rangle$, defined in what follows with natural orbitals computed in NRG. The product goes from $-N/2$ to $-M/2 - 1$, which corresponds (with the label of orbitals defined above) to the $(N - M)/2$ first eigenvalues of Q that are the closest to 1. The remaining $(N - M)/2$ uncorrelated orbitals have an occupancy $\lambda_n \simeq 0$ and do not appear in the determinant $|\Psi_0\rangle$, they are treated as exactly empty.

The full wave function considering the total number N of natural orbitals thus reads:

$$|\Psi_{\text{few}}\rangle = \sum_{\{N_n=0,1\}} \Psi(N_{-\frac{M}{2}}, \ldots, N_{\frac{M}{2}}) \prod_{n=-\frac{M}{2}}^{\frac{M}{2}} [q_n^\dagger]^{N_n} |\Psi_0\rangle, \tag{3.13}$$

with $N_n = 0, 1$ the occupancy of the correlated orbital $q_n^\dagger$, where the summation is restricted to occupancies such that $\sum_{n=-M/2}^{M/2} N_n = M/2$ to be at half-filling (where the ground state obeys PH symmetry), and $\Psi(N_{-\frac{M}{2}}, \ldots, N_{\frac{M}{2}})$ are expansion coefficients of the many-body wave function of the correlated sector. Such an ansatz is common in quantum chemistry calculations (known as the complete active space (CAS) description), where core orbitals are frozen and only a part of valence orbitals are taken as active to approximate electronic correlations in molecules [4, 7, 22–24].

The Hamiltonian of the IRLM can be re-expressed with natural orbitals to obtain the full wave function. It this thus transformed (without approximation here) by the orthogonal matrix D that diagonalizes Q as:

$$\mathcal{H}_{\text{IRLM}} = \mathcal{H}_{\text{corr}} + \mathcal{H}_{\text{uncorr}} + \mathcal{H}_{\text{mix}}, \tag{3.14}$$

with each part defined by the nature of orbitals involved, that either belong only to the correlated sector for $\mathcal{H}_{\text{corr}}$, and the uncorrelated sector for $\mathcal{H}_{\text{uncorr}}$. $\mathcal{H}_{\text{mix}}$ concerns terms with orbitals appearing in both sectors. From now, the following convention to write tensors will be used to lighten the notation: a superscript U^a corresponds to the matrix element linked to a creation operator $q_a^\dagger$, and a subscript U_a is linked to an annihilation operator q_a. The superscripts and subscripts stacks as more than one operator are concerned. The three terms in the Hamiltonian are thus written explicitly:

$$\begin{aligned}
\mathcal{H}_{\text{corr}} &= \sum_{n,m} t^n_m \, q_n^\dagger q_m + \sum_{n,m,p,q} U^{nm}_{pq} \, q_n^\dagger q_m^\dagger q_p q_q \\
\mathcal{H}_{\text{mix}} &= \mathcal{H}_{\text{odd}} + \sum_{n,m} \sum_{\alpha,\beta} \Big(U^{n\alpha}_{m\beta} \, q_n^\dagger q_\alpha^\dagger q_m q_\beta + U^{n\alpha}_{\beta m} \, q_n^\dagger q_\alpha^\dagger q_\beta q_m \\
&\qquad + U^{\alpha n}_{m\beta} \, q_\alpha^\dagger q_n^\dagger q_m q_\beta + U^{\alpha n}_{\beta m} \, q_\alpha^\dagger q_n^\dagger q_\beta q_m \Big) \\
\mathcal{H}_{\text{uncorr}} &= \sum_{\alpha,\beta} t^\alpha_\beta \, q_\alpha^\dagger q_\beta + \sum_{\alpha,\beta,\gamma,\delta} U^{\alpha\beta}_{\gamma\delta} \, q_\alpha^\dagger q_\beta^\dagger q_\gamma q_\delta .
\end{aligned} \tag{3.15}$$

The elements of the tensor U^{ab}_{cd} are defined as $U^{ab}_{cd} = U D^\dagger_{-1a} D^\dagger_{0b} D_{0c} D_{-1d}$ corresponding to the form of the interaction of the IRLM, where $d^\dagger d c_0^\dagger c_0$ has been transformed to $d^\dagger c_0^\dagger c_0 d$ for a matter of conventions. $\mathcal{H}_{\text{odd}}$ contains terms with an odd number of particles in the correlated or uncorrelated sectors, and these terms exactly vanish when the Hamiltonian is evaluated in a state written as Eq. (3.13). Correlated orbitals are labeled with roman indices $n, m \in [-M/2, M/2]$ and uncorrelated ones with greek indices $\alpha, \beta \in [-N/2, -M/2-1] \cup [M/2+1, N/2]$.

In this basis, the uncorrelated part of the state $|\Psi_0\rangle$ is already known, and we can get rid of frozen degrees of freedom q_α such that the resulting few-body Hamiltonian to solve can be reduced to:

$$\begin{aligned}
\mathcal{H}_{\text{few}} &= \langle \Psi_0 | \, \mathcal{H}_{\text{IRLM}} \, | \Psi_0 \rangle \\
&= \sum_{n,m} t^n_m q_n^\dagger q_m + \sum_{n,m,p,q} U^{nm}_{pq} \, q_n^\dagger q_m^\dagger q_p q_q \\
&\quad + \sum_{nm} \sum_{\alpha} q_n^\dagger q_m n_\alpha \Big(- U^{n\alpha}_{m\alpha} + U^{n\alpha}_{\alpha m} + U^{\alpha n}_{m\alpha} - U^{\alpha n}_{\alpha m} \Big) \\
&\quad + \sum_{\alpha} t^\alpha_\alpha n_\alpha + \sum_{\alpha \neq \beta} (U^{\alpha\beta}_{\beta\alpha} - U^{\alpha\beta}_{\alpha\beta}) n_\alpha n_\beta ,
\end{aligned} \tag{3.16}$$

with $n_\alpha = \langle\Psi_0| q_\alpha^\dagger q_\alpha |\Psi_0\rangle$. Non-zero terms in $\langle\Psi_0| \mathcal{H}_{\text{mix}} |\Psi_0\rangle$ acts as Hartree terms renormalizing the hopping between correlated orbitals, and $\langle\Psi_0| \mathcal{H}_{\text{uncorr}} |\Psi_0\rangle$ is just a global shift in energy for the correlated sector, since no correlated orbitals come from this term (see last line of Eq. (3.16)). Note that no approximations were made on the Hamiltonian part, but only on the form of the wave function $|\Psi_{\text{few}}\rangle$ in Eq. (3.13).

The optimal few body wave function $\Psi(N_{-\frac{M}{2}}, \ldots, N_{\frac{M}{2}})$ can be found by exact diagonalization of the few-body Hamiltonian $\mathcal{H}_{\text{few}}$. Contrary to the NRG, which only gives the state in the truncated many-body basis, this approximated wave function is expressed with one-body orbitals that can be related to the physical degrees of freedom c_i and is thus more practical to use. Remember that in the NRG, any operator must be defined in parallel to the whole flow of the Hamiltonian, that makes the computation of observables a bit more involved. The only relevant parameter of the few-body approximation of the wave function is M, that counts the number of correlated orbitals. In the limit $M \to N$, the exact wave function is recovered. In the non-interacting case $U = 0$, the separation of sectors is exact for any even value of M.

The accuracy of the few-body ansatz for different degrees of correlation is tested by comparing the *variational* ground state energy $E_{\text{few}} = \langle\Psi_{\text{few}}|\mathcal{H}_{\text{IRLM}}|\Psi_{\text{few}}\rangle$ to the true ground state energy E_{NRG} computed with NRG, and the results are plotted in Fig. 3.7. Natural orbitals defining $|\Psi_{\text{few}}\rangle$ are computed with the NRG. Each computation at finite M is still in the thermodynamic limit, since every $N - M$ orbitals are fully accounted in the wave function through $|\Psi_0\rangle$. We find that the few-body energy converges exponentially to the true ground state as M is increased, as anticipated from the structure of the correlation spectrum. While $M \ll N$ orbitals are considered, the ground state is at half-filling such that only $M/2$ particles bear correlations in the system. As U is pushed deeper in the correlated regime, the convergence becomes slower and the ansatz loses accuracy. The convergence of the energy can be related to the occupancy of the first pair of frozen orbitals: $E_{\text{few}} - E_{\text{NRG}} \simeq \lambda_{M/2+1}$. Hence,

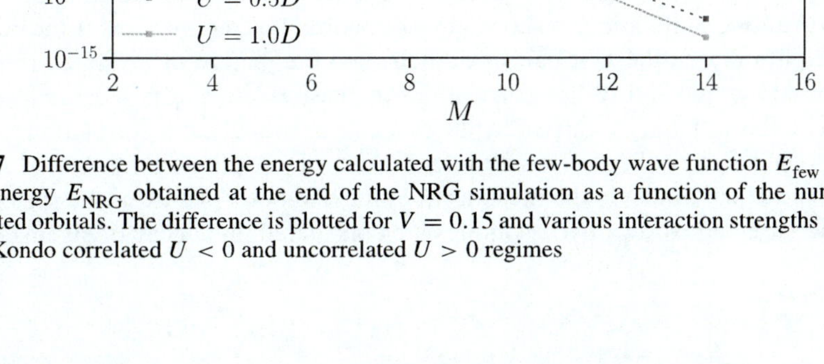

Fig. 3.7 Difference between the energy calculated with the few-body wave function E_{few} and the exact energy E_{NRG} obtained at the end of the NRG simulation as a function of the number of correlated orbitals. The difference is plotted for $V = 0.15$ and various interaction strengths U, both in the Kondo correlated $U < 0$ and uncorrelated $U > 0$ regimes

even for $U = -0.5$ ($T_K \simeq 10^{-3}$), 5 particles in $M = 10$ correlated orbitals suffice to converge the energy at 6 digits. This clearly vindicates that ground states of quantum impurity problems are few-body, unless they are tuned exponentially close to a phase transition. Equivalent results are found with the SIAM with a doubled value of M to account for the spin degrees of freedom.

3.6 One-Body Picture and Spatial Dispersion of Natural Orbitals

It was demonstrated in this chapter that only few orbitals were participating in correlations spreading in the host surrounding a quantum impurity, and that this number of orbitals is increasing when correlations become stronger in the system. In Refs. [15, 21], it was proposed an even stronger statement, suggesting that only a single orbital is responsible for the screening process of the magnetic moment carried by the impurity, in the strong and weak coupling limit of the Kondo model. While the strong coupling limit $J \to \infty$ of the Kondo model is trivial (a bound state is formed with the first site of the bath and is thus decoupled from the environment), the weak coupling regime $J \to 0$ is strongly correlated and this naive picture seems too restrictive.

In the study of Ref. [21], real space lattices are simulated with DMRG, and systems up to hundreds of sites are investigated. The authors study finite size effects on the Kondo cloud: when J is lowered from the bandwidth D, it is first found that the picture of a single orbital becomes worse and worse, but becomes accurate again until it appears to be exact at $J \to 0^+$. They work at fixed system size, and what they truly observe is the following: when $L_K \sim 1/T_K$ (which increases when J decreases) is lower than the system size, spatial Kondo correlations can extend further in the system, and more and more orbitals are involved in these correlations, destroying the one-body picture. However when L_K is reached, correlations are cut off and the system becomes simple again if $L_K \gg L$.

In renormalization group studies, the physical length scale L gets larger and larger with the iterations to reach low energy properties. However, each length scale has to be considered with respect to the emerging scale of the system, that is the Kondo length. The equivalent system that is considered in DMRG studies of fixed $L \sim 100 - 500$ when $L_K \to \infty$ is similar to having a handful of sites only for a realistic L_K, such that the picture of a single coupled orbital is trivial.

In Fig. 3.8, the spatial extent of the first most correlated orbitals computed by converged NRG simulations is shown. In the uncorrelated regime $U > 0$, the impurity has weight only on the first pair of orbitals, and the picture of a single orbital can be approximately valid. When correlations increase ($U < 0$), the screening process appears to be still mostly carried by the first pair of orbitals, but correlations carried by the others can not be neglected for an accurate description of the ground state.

To study quantitatively the breakdown of the one-body picture as the system size L is increased above L_K, two quantities are considered: the decay of eigenvalues of

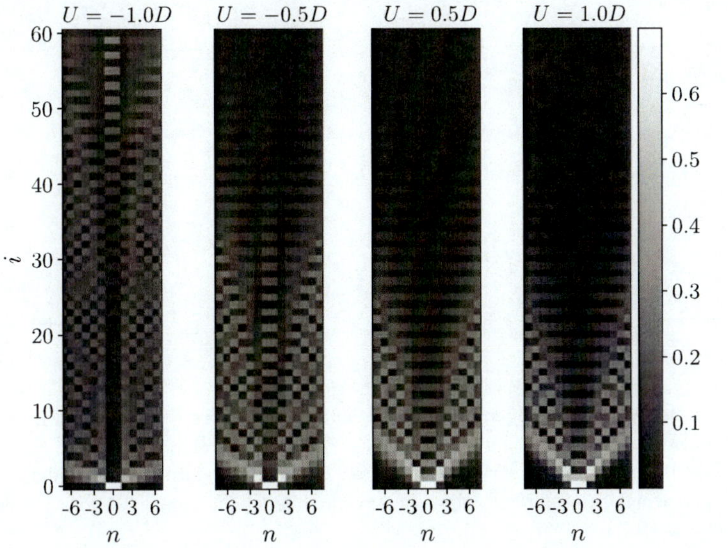

Fig. 3.8 Spatial dispersion of the 7 most correlated pairs of PH conjugated natural orbitals q_n for four interaction strengths and for the same parameters as in previous figures. It is defined as the absolute value of elements of the orthogonal matrix D_{in} at each site i and orbital n. The spatial extent is cut at site $i = 60$ for better visibility around the impurity, but the numerical simulations end at $N = 110$ for the results to be converged. In the uncorrelated regime ($U > 0$), natural orbitals are mostly Wannier orbitals with some slight amplitude along the chain. When correlations gets stronger ($U < 0$), the impurity level is strongly hybridized with Wannier orbitals far from the impurity at distances $\sim L_{\rm K}$, and each natural orbital spreads throughout the entire system (impurity included)

Q and the *correlation entropy* $S_{\rm corr}$, both against L. The correlation entropy that is studied here is defined for every natural orbital except the mostly correlated pair in our system as:

$$S_{\rm corr} = \sum_{n=-N/2}^{-3} \left(n_n \ln(n_n) + n_{-n} \ln(n_{-n}) \right) . \tag{3.17}$$

If $S_{\rm corr} = 0$, only the mostly correlated pair bears correlations, and the one-body picture is correct. In the left panel of Fig. 3.9, the decay rate x of the SIAM is plotted against the system size L for $U = 55\Gamma$. In this case, $T_{\rm K} \simeq 5 \times 10^{-10}$ so that NRG simulations can reach both $L \ll L_{\rm K}$ and $L \gg L_{\rm K}$. L is defined during the RG flow, each step n defining a length scale $L = 1/\Lambda^{-n/2}$. The decay rate is decreasing when $L \to L_{\rm K}$ and then saturates at a minimum. Correlations thus decay much faster when the system size is exponentially small compared to the Kondo length. $S_{\rm corr}$ is plotted in the right panel of Fig. 3.9, and shows the same behavior as x: it saturates at a maximum value after $L_{\rm K}$ is reached, and decreases as L goes to 0. Strictly speaking, the one-body perspective is true when $L \to 0$ or $L_{\rm K} \to \infty$. However, these statements are trivial or unrealistic, such that the few-body perspective that we

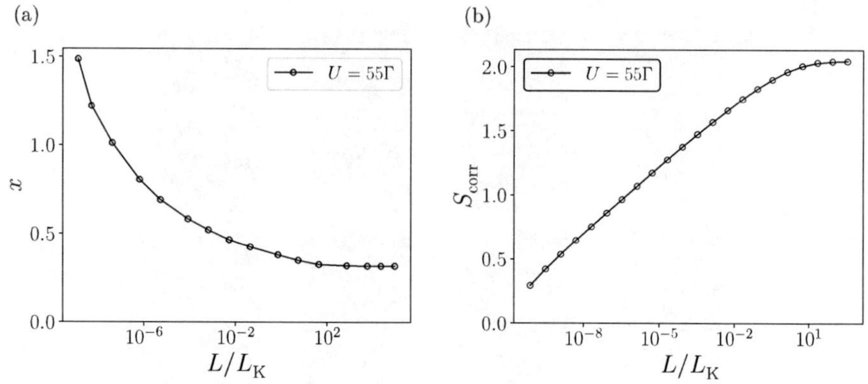

Fig. 3.9 **a** Decay rate x of eigenvalues of the correlation matrix Q of the SIAM, for $U = 55\Gamma$, $\Gamma = 10^{-2}$ and $\Lambda = 2.0$, against the rescaled size L/L_{K}. **b** Correlation entropy S_{corr} as a function of the rescaled size, for the same parameters as (**a**). It increases logarithmically with the system size until L_{K} is reached then saturates, indicating that sites added far behind L_{K} does not participate to correlations at all

proposed in this chapter is more general and provides a flexible description of the system, whose accuracy can be tuned with the number of correlated orbitals M.

A quantitative study of correlations in quantum impurity systems using natural orbitals has been presented in this chapter. As was expected from previous studies, an exponential decay of the occupancies of natural orbitals has been found, which has been used to propose an ansatz for the wave function reducing the full-many body problem to a few-body system of $M \ll N$ correlated orbitals. Nevertheless, this few-body wave function relies on the knowledge of natural orbitals, that are eigenvectors of the one-body density matrix Q of the problem, that was computed through NRG simulations. NRG requires a discretization of the bath degrees of freedom, and does not take advantage of the hierarchical structure of correlations in these systems. It would be desirable to find an NRG-independent method to compute natural orbitals, in order to benefit from the practical form of the few-body wave function without the disadvantages of the logarithmic discretization.

During this thesis, a numerical method based on renormalization group ideas has been developed to find an optimized set of M natural orbitals independently, and is presented in the following Chap. 4. This method scales polynomially as a function of the number of sites in the system, so that real space calculations with large system sizes (up to some tens of thousands of sites) become possible with a reasonable computating power and a controlled accuracy.

References

1. Löowdin P-O (1955) Quantum theory of many-particle systems. I. Physical interpretations by means of density matrices, natural spin-orbitals, and convergence problems in the method of configurational interaction. Phys Rev 97:1474–1489
2. Davidson ER (1972) Natural orbitals. Academic, pp 235–266
3. Roos BO, Taylor PR, Sigbahn PEM (1980) A complete active space SCF method (CASSCF) using a density matrix formulated super-CI approach. Chem Phys 48:157–173. ISSN: 0301-0104
4. Andersson K, Malmqvist P-Å, Roos BO (1992) Second-order perturbation theory with a complete active space self-consistent field reference function. J Chem Phys 96:1218–1226. ISSN: 0021-9606
5. David Sherrill C, Schaefer HF (1999) The configuration interaction method: advances in highly correlated approaches. Academic, pp 143–269
6. Held K et al (2006) Realistic investigations of correlated electron systems with LDA + DMFT. Physica status solidi (b) 243:2599–2631
7. Kotliar G et al (2006) Electronic structure calculations with dynamical mean-field theory. Rev Mod Phys 78:865–951
8. Zgid D, Chan GK-L (2011) Dynamical mean-field theory from a quantum chemical perspective. J Chem Phys 134:094115. ISSN: 0021-9606, 1089-7690
9. Zgid D, Gull E, Chan GK-L (2012) Truncated configuration interaction expansions as solvers for correlated quantum impurity models and dynamical mean-field theory. Phys Rev B 86:165128. ISSN: 1098-0121, 1550-235X
10. Lin C, Demkov AA (2013) Efficient variational approach to the impurity problem and its application to the dynamical mean-field theory. Phys Rev B 88:035123. ISSN: 1098-0121, 1550-235X
11. Lu Y, Höppner M, Gunnarsson O, Haverkort MW (2014) Efficient real-frequency solver for dynamical mean-field theory. Phys Rev B 90:085102. ISSN: 1098-0121, 1550-235X
12. He R-Q, Lu Z-Y (2014) Quantum renormalization groups based on natural orbitals. (2014). arXiv:1306.3662 [cond-mat, physics:physics, physics:quant-ph].
13. Yang C, Feiguin AE (2017) Unveiling the internal entanglement structure of the Kondo singlet (2017). arXiv:1703.02383 [cond-mat].
14. Go A, Millis AJ (2017) Adaptively truncated Hilbert space based impurity solver for dynamical mean-field theory. Phys Rev B 96:085139. ISSN: 2469-9950, 2469-9969
15. Zheng R, He R-Q, Lu Z-Y (2018) Natural orbitals renormalization group approach to a Kondo singlet. arXiv:1803.03072 [cond-mat].
16. Lu Y, Cao X, Hansmann P, Haverkort MW (2019) Natural-orbital impurity solver and projection approach for Green's functions. Phys Rev B 100:115134. ISSN: 2469-9950, 2469-9969
17. Debertolis M, Florens S, Snyman I (2021) Few-body nature of Kondo correlated ground states. Phys Rev B 103:235166
18. Kitatani M, Sakai S, Arita R (2021) Natural orbital impurity solver for real-frequency properties at finite temperature. arXiv:2107.06517 [cond-mat].
19. Debertolis M, Snyman I, Florens S (2022) Simulating realistic screening clouds around quantum impurities: role of spatial anisotropy and disorder. Phys Rev B 106:125115
20. Bravyi S, Gosset D (2017) Complexity of quantum impurity problems. Commun Math Phys 356:451-500. ISSN: 1432-0916
21. Yang C, Feiguin AE (2017) Unveiling the internal entanglement structure of the Kondo singlet. arXiv:1703.02383 [cond-mat]
22. Siegbahn PE-M, Almlöf J, Heiberg A, Roos BO (1981) The complete active space SCF (CASSCF) method in a Newton-Raphson formulation with application to the HNO molecule. J Chem Phys 74:2384–2396
23. Helgaker T, Jorgensen P, Olsen J (2000) Molecular electronic-structure theory. Wiley
24. Olsen J (2011) The CASSCF method: a perspective and commentary. Int J Quant Chem 111:3267–3272

Chapter 4
Recursive Generation of Natural Orbitals

Check for updates

4.1 Motivations

The exponential decay of the one-body density matrix eigenvalues that was investigated in Chap. 3 opens new roads to tackle quantum impurity problems. The strong hierarchical structure of correlations allows to consider different kind of relevant descriptions, the few-body ansatz for the wave function presented previously in the text being one of them. Another description relies on wave functions defined as a linear superposition of non-orthogonal Gaussian states [1–3]. The weight of each Gaussian state in the wave function is determined by minimization over the manifold of parameters constituting each state. The number of Gaussian states is expected to be small when the underlying orbitals are natural orbitals (at least when T_K is reasonably large) for quantum impurity systems, which makes the strength of this approximation. However, as the size of the system increases, the manifold of parameters grows and minimization methods lose efficiency. The few-body ansatz on its side, which relies on a strict separation of a set of correlated and uncorrelated orbitals, the latter containing the advantage to work in the one-body Hilbert space, will be the building block of the method presented in this chapter. This new method aims to find an optimal set of correlated orbital, from which the few-body wave function can be defined to compute any observable in the ground state. The uncorrelated space is defined up to a rotation, and is simply taken as the orthogonal complement of the correlated space, such that the scaling of the state of interest is not extensive with the system size.

Equipped with this efficient description, it is wise to think about how this approach can help address issues that were 0difficult in the past. Studies of Kondo-correlations in disordered host, that were presented in Chap. 2, clearly lack of a microscopic picture of spatial correlations, that would require efficient methods to perform statistics. The Kondo cloud, that was also introduced in this chapter is at the core of quantum impurity problems, and has also proven expensive to compute numerically due to limitations of existing methods. A microscopic investigation of this spatial cloud in

M. Debertolis, *Quantum Impurity Problems in the Framework of Natural Orbitals*, Springer Theses, https://doi.org/10.1007/978-3-031-47233-6_4

higher dimension is also missing from the literature, that usually considers spherical s-wave scattering around impurities reducing the problem to one dimension. During this thesis a tailored quantum impurity solver that aims to address all these questions has been developed, and it is presented in this chapter.

As was stressed earlier, an efficient resolution of impurity problems is required for DMFT calculations, and various algorithms based on natural orbitals have been proposed in the past decade [4–8]. Most of these algorithms aim to compute dynamical self-energies required by the self-consistent loop of DMFT. These methods are not adapted when the wave function is sought, so that an independent generation of a set of optimized correlated orbitals beyond Hartree-Fock calculations were proposed aside, using minimization procedures, or iterative schemes [9–11]. However, they are based on different ansatz or use global minimization steps that are usually less efficient than a fully iterative procedure. The algorithm that will be presented here recursively generates improved guesses for the basis of natural orbitals by incorporating new physical orbitals to the system, and discarding unrelevant ones until convergence (a similar implementation was proposed for quantum chemistry applications [9]).

The efficiency of this recursive generation of natural orbitals (RGNO) algorithm will be investigated for different degrees of correlations in the system, and a confrontation to NRG results will be presented to prove its accurate description of the ground state. Then, the Wilson chain can be abandoned in favor of a real space tight-binding chain, that will reveal the true power of the algorithm to resolve real space structure on large-scale systems, within which the computation of spatially non-local observables such as the Kondo screening cloud is less demanding. In Chap. 5, a microscopic study of Kondo-correlations in the IRLM in presence of charge disorder using the new algorithm will be discussed.

4.2 Recursive Generation of Natural Orbitals (RGNO)

4.2.1 General Idea

The RGNO is a non-perturbative method based on the few-body ansatz for the wave function (3.13), that can be written in a simpler way (the notation of Chap. 3 for orbitals and tensors are used):

$$|\Psi\rangle = \prod_{\alpha=-N/2}^{-M/2-1} q_\alpha^\dagger \sum_S \Psi_S \, |S\rangle \, , \tag{4.1}$$

where $|S\rangle$ is a Slater determinant involving a set of correlated orbitals $q_{n=-M/2,\ldots,M/2}$, whose weight in the ground state wave function is given by Ψ_S. These determinants describe a number N_{corr} of fermions occupying the M active orbitals, with $N_{\text{corr}} = M/2$ in the case of half-filling. The uncorrelated orbitals q_α are orthogonal to the correlated ones, and those with an energy below the Fermi energy $E_{\text{F}} = 0$ are occupied in the ground state. Thus, uncorrelated orbitals separate into an occupied and an unoccupied sector, that are PH conjugated when the symmetry is not broken. For a given choice of orbitals in the three sectors (occupied, correlated, unoccupied), the coefficients Ψ_S are determined by minimizing the ground state energy, defined as the expectation value of the effective correlated Hamiltonian defined as (3.16) in the state $|\Psi\rangle$. In the previous study, a prior computation of NOs with NRG simulations allowed to work with the optimal set of orbitals for the three sectors, and the accuracy of the ansatz was only dependent on the size of the correlated sector M. Here, the quality of the ansatz hinges firstly on the choice of orbitals spanning each three orbital space. Unfortunately, the one-body matrix through which NOs are defined is unknown, as it needs to be calculated from the exact ground state.

The method aims to find without recourse to NRG the optimal set of M correlated orbitals, which are expected to be close to the M most correlated orbitals of the true correlation matrix Q.

The algorithm starts by choosing an initial reparation of a given set of orbitals in the three different sectors. This first guess of orbitals has no chance to be the right one, and needs to be improved. For this purpose, the correlated space is expanded by adding two well-chosen orbitals belonging to other sectors, that contains now $M + 2$ orbitals. In this expanded basis, the ground state $|\Psi^{\mathrm{add}}\rangle$ of the effective enlarged Hamiltonian can be computed, from which the correlation matrix containing the $M + 2$ correlated orbitals can be computed: $Q_{ij} = \langle\Psi^{\mathrm{add}}|q_i^\dagger q_j|\Psi^{\mathrm{add}}\rangle$. As the two added orbitals are not eigenstates of Q, the new set of eigenvectors is different from the last one. Then, the two eigenvectors of the enlarged Q with eigenvalues closest to 0 and 1 are moved to the uncorrelated space, thus leaving a new state of M correlated orbital as the improved new guess. At this point, the procedure of adding and throwing orbitals is repeated until convergence to a fixed point of $|\Psi\rangle$.

A simple verification for the convergence of the state is the decrease of the ground state energy of the effective Hamiltonian for each improvement of correlated orbitals. The converged state is benchmarked against exact NRG results, and it appears that the state is almost as accurate as state (4.1) with the exact natural orbitals and shows the same exponential convergence with the number of correlated orbitals M. This iterative procedure is similar to an RG for the parameters of the Hamiltonian in the following sense: the matrix elements of the effective Hamiltonian are defined by the transformation matrix D from Wannier to natural orbitals, which changes at each step of the flow until it reaches a fixed point. In the following, the implementation of the RGNO is detailed for the IRLM Hamiltonian on the Wilson chain (3.11) in order to be compared to the NRG simulations of this model.

4.2.2 *Initial Guess of Correlated Orbitals*

The first step is to choose a initial complete set of orbitals to start the flow of orbitals, and to distribute them between the three sectors. As for every iterative procedure, the closer the initial guess to the solution the faster the convergence. An optimal initial guess is the Hartree-Fock set of orbitals, defined through the resolutions of self-consistent equations similar to Eqs. (2.21) that were introduced in Chap. 2. Building these orbitals adds a slight computational effort to the global procedure yet the guess proposed hereafter is cheaper to implement and provides comparable results.

In the previous study, the spatial dispersion of the exact natural orbitals in Fig. 3.8 revealed that, generally, the impurity level is not diluted too much into other orbitals, such that the interaction stays local at zeroth order. This indicates that Hartree-Fock equations do not renormalize a lot the parameters of the system, and that the bare decoupling of the interaction is not far from the mean field solution. Hence, an initial guess can be the following: the first M sites (Wannier orbitals) are taken as the correlated space, and the remaining $N - M$ will constitute the uncorrelated space.

To differentiate occupied and unoccupied (PH conjugated) orbitals, the $N - M$ sector is diagonalized (there is no interaction between these orbitals, the diagonalization is done in the Hilbert space): if $|P_{\text{unc}}\rangle$ is the projector into this subspace, the projected Hamiltonian $\mathcal{H}_{\text{IRLM}}|P_{\text{unc}}\rangle\langle P_{\text{unc}}|$ is considered. Each of its eigenvectors with energy $\epsilon_n < 0$ is occupied, and the remaining ones are unoccupied. This works well because this spectrum is nearly degenerate with respect to the one of the one-body density matrix Q. We could also choose to take the eigenvectors of the free problem $U = 0$ as a starting point, including the impurity in every sector. Actually, it depends if the expected natural orbitals are localized in space (hence the first solution may be better), or in energy (the second one is thus more appropriate). The first choice of orbitals is depicted in Fig. 4.1.

With the chosen initial guess, orbitals will now be improved by the iterative expansion and reduction of the correlated space. Note that if the initial guess is too far from

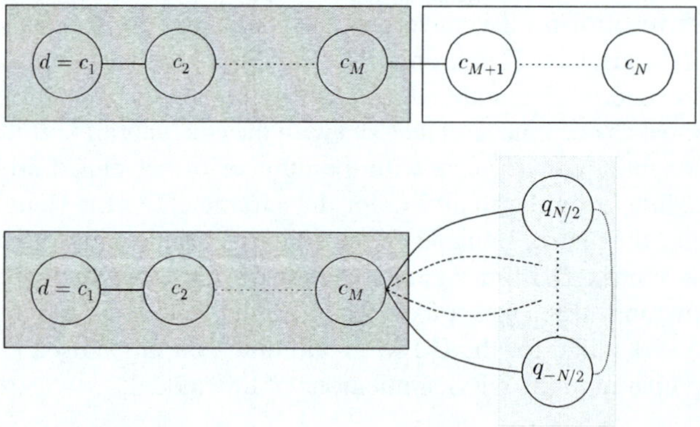

Fig. 4.1 Initialization procedure of the RGNO. The first M Wannier orbitals (in dark grey) are chosen to describe the correlated space, whereas the uncorrelated part is defined from the $N - M$ remaining orbitals. The occupied and unoccupied sectors of the uncorrelated space are determined through the diagonalization of the free Hamiltonian projected on the uncorrelated space. Eigenvectors q_α, with $|\alpha| \in [M + 1, N/2]$ are paired according to their relative energy to the chemical potential, that is set to zero here. Hence, $q_{-\beta}$ is paired with q_β, and q_{M+1} has the closest energy to the Fermi energy

the solution, the algorithm may be stuck in a local minimum and yield inconsistent results. The two presented initializations are the ones that lead to drastically faster convergence of the algorithm.

4.2.3 Iterative Diagonalization

The iterative diagonalization aims to improve the initial separation of orbitals in the three sectors. The correlated space will be successively enlarged by two uncorrelated orbitals, one from the occupied sector and one from the unoccupied one. Orbitals that are most likely to participate in correlations are added first: they correspond to the ones with energies the closest to the Fermi energy, namely ϵ_n the closest to 0. Uncorrelated orbitals are sorted in ascending order, such that $q_{-N/2}$ (resp. $q_{N/2}$) is the furthest in energy from 0 and $q_{-M/2-1}$ (resp. $q_{M/2+1}$) the closest. Hence, the pair $(q_{-M/2-1}, q_{M/2+1})$ is added first, and $(q_{-N/2}, q_{N/2})$ will be the last added pair of orbitals to the flow. The enlarged Hamiltonian is defined as in Eq. (3.16) for $M+2$ orbitals, the $(N-M-2)/2$ frozen orbitals being treated in mean-field. The Hamiltonian is diagonalized to find $|\Psi^{\text{add}}\rangle$, and the corresponding Q matrix is computed. Then, the two least correlated orbitals are thrown (downgraded to uncorrelated) to keep a space of fixed size, and the procedure is repeated. The improvement procedure is sketched in the top panel of Fig. 4.2. As orbitals are thrown, they still participate to the renormalization of the couplings between correlated orbitals, such that the set of orbitals stays complete at every moment of the flow. The separation of sectors for a given iteration n is shown in the bottom panel of Fig. 4.2.

For each definition of the Q matrix, the exact diagonalization of the effective enlarged Hamiltonian is done in the Fock space of size 2^{M+2}. However, the Hamiltonian is very sparse, and the absolute number of states to consider is in practice way smaller than 2^{M+2}. The Hamiltonian conserves the number of particles, such that it can be diagonalized by blocks of fixed number of particles, which already reduces a lot the size of the corresponding matrix. The Hamiltonian (3.11) is particle-hole symmetric, and it would be beneficial to implement this symmetry in the exact diagonalization. However, at each step, the Hamiltonian that is considered is already projected on the uncorrelated part of the wave function, and it appears that this projection explicitly breaks PH symmetry (see Appendix C). Forcing the PH symmetry in the exact diagonalization has been tested, and lead to numerical instabilities along the flow that prevents the convergence of the state. A solution could be to symmetrize the ansatz by taking $|\Psi\rangle_{\text{sym}} = |\Psi\rangle + P|\Psi\rangle$ (with the PH conjugation operator P defined as in Eq. (3.9)), but this definition is less tractable and not essential in this study, so that the slight symmetry breaking is tolerated. However, even without the PH symmetry, the Hamiltonian is sparse enough to deal with at most 14 correlated orbitals, which is more than enough for the following studies. The detailed implementation of the sparse Hamiltonian for an optimized exact diagonalization is presented in Appendix D.

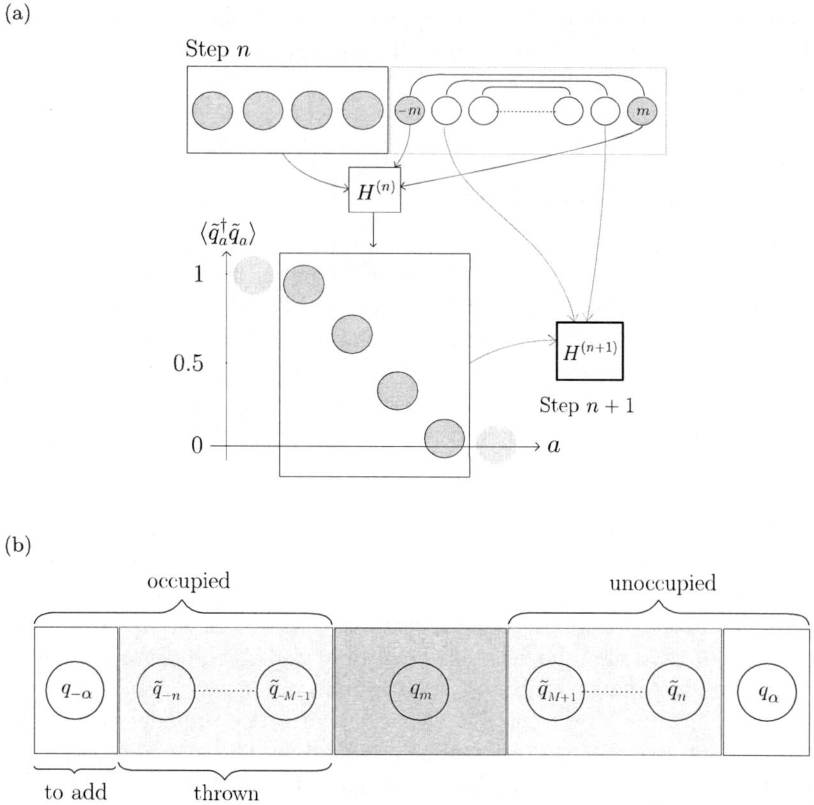

Fig. 4.2 **a** Transformation of the Hamiltonian from step n to $n+1$. The correlated sector of $H^{(n)}$ is enlarged by the uncorrelated pair of orbitals (q_m, q_{-m}), with $m = M/2 + 1 + n$ the nth added pair or orbitals. The corresponding matrix Q is diagonalized in the enlarged ground state of $H^{(n)}$, defining a new set of correlated orbitals $\tilde{q}_i$. The two least correlated orbitals (darker circles) are moved to the uncorrelated space, such that the Hamiltonian $H^{(n+1)}$ to be enlarged in the next step $n+1$ is of size M. **b** Repartition of orbitals at a given step n of the flow between the three sectors. Thrown orbitals and uncorrelated ones defined in the initialization of the flow are considered in the same way, the basis of orbital being complete at each step of the flow

During this procedure, we chose to add orbitals by pairs in the correlated space, which is the minimum required to respect at best the PH symmetry (even if it is inherently slightly broken by the ansatz for the wave function). It has been tested to add orbitals four by four instead, but the convergence speed remains the same while the intermediate diagonalization becomes heavier (the size increases to 2^{M+4} states).

4.2.4 Sweep Until Convergence

The iterative diagonalization described previously is repeated until each pair of uncorrelated orbitals has been added to the flow, and thus constitutes a sweep. At the end of the first sweep, the corresponding correlated space has improved, but nothing guarantees that it is optimal. In order to reach full convergence of NOs for a given size M of the correlated space, the NOs obtained at the end of a sweep are taken as the new guess for the next one, until the final fixed point is reached. The uncorrelated sector is thus diagonalized once per sweep, and the $(N - M)/2$ steps of adding / throwing repeated. To monitor the convergence of NOs, the ground state energy or any observable that needs to be computed can be computed at the end of each sweep, and in practice for clean systems, 10 to 20 sweeps are enough to reach full accuracy.

The RGNO method presented above was found to be the optimal one in any case, and only the accuracy of the first set of orbitals can accelerate the convergence. The scaling of the algorithm is at most $(N - M)^3$ ($\simeq N^3$ for $N \gg M$) for the diagonalization of the uncorrelated space, and $(N - M)/2$ ($\simeq N/2$) times 2^{M+2} for each exact diagonalization of the correlated space at each step, times the number of sweeps that is independent of N. The total scaling is thus of order $\mathcal{O}\left(N^4 2^M\right)$, which is polynomial in the system size N, for an exponential precision (that will be discussed just after) against M. As long as M is small enough ($<$ 10 to 14 without the use of symmetries), the procedure is relatively fast for large systems up to several thousands of sites.

At the beginning of each sweep, the uncorrelated sector must be diagonalized to determine in which order will orbitals be added. When N gets too large, the diagonalization in N^3 can be the bottleneck of the algorithm, such that it could be interesting to avoid this step at each sweep. It has been tested that using the same set of uncorrelated orbitals but sorted in reverse order (the last thrown will becomes the first added) leads to results with only slightly lower precision, but the time of calculation saved by this trick makes it profitable for large enough N.

4.3 Benchmark on the Wilson Chain

The accuracy of the wave function produced with the RGNO is tested against converged NRG results. The same grid of discretization is taken for both algorithms, and the NRG is ensured to be converged, following the same verifications as in Chap. 3. The parameters used in the NRG simulation for this study, for which the thermodynamic limit is reached, are $\Lambda = 2$, $N = 110$ sites and a total number of kept states $N_{\text{kept}} = 1500$. The algorithm is tested in both the weakly correlated $U > 0$ and strongly correlated $U < 0$ regimes of the IRLM to put to test its robustness to strong correlations.

In the left panel of Fig. 4.3, the ground state energies obtained by RGNO is compared to the one computed in NRG, for different interaction strength yielding different

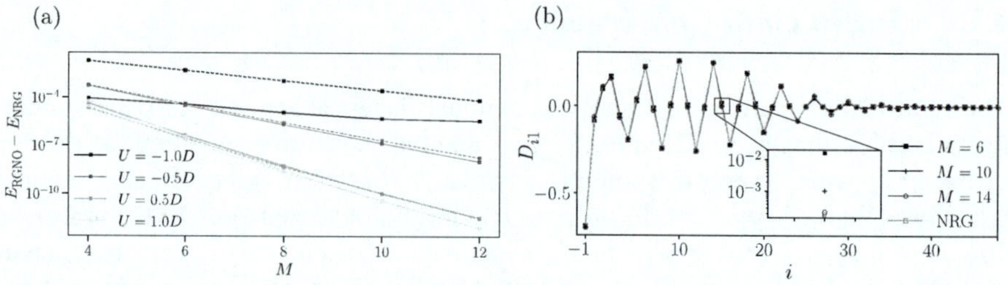

Fig. 4.3 **a** Difference between the ground state energy of the IRLM computed with the RGNO and NRG against the number of correlated orbitals M for different values of the interaction U. The calculation is performed within both algorithm for $\Lambda = 2.0$, $V = 0.15$, $N = 110$ sites. The dashed lines corresponds to the convergence of the few-body ansatz with nearly exact NOs computed in NRG, plotted in Fig. 3.7. The curve at $U = -1.0$ is not converged by the RGNO due to the slight symmetry breaking of the ansatz. **b** Amplitude D_{i1} of the most correlated orbital q_1 along each site of the chain i (with $i = -1 \equiv d$) computed with both RGNO and NRG. The chain is cut at $i = 48$ for readability, the amplitude being almost zero after $i \sim 30$. The inset shows the convergence of the RGNO orbital to the NRG result for $D_{15,1}$

T_K. The algorithm converges exponentially with the number of correlated orbitals M to the true ground state energy, that is typical for the exponentially decaying spectrum of Q eigenvalues (see the convergence of the few-body ansatz in Fig. 3.7). For $U = -0.5$, the approximate ground state energy is converged up to 5 digits to the true ground state, that is well below the Kondo temperature for these parameters $T_K \simeq 3 \times 10^{-3}$. Kondo correlations are ensured to be fully captured for $M \geq 6$. When correlations are weaker ($U > 0$ or $U < U_c$), the energy is easily converged with a precision better than 10^{-8} for only $M = 8$ orbitals.

In the non-interacting case $U = 0$, it is recalled that the construction of the ansatz is exact such that the algorithm finds the exact ground state wave function (that is not ensured for every method suited for many-body regimes). Results for $U = -1.0$ does not follow the exponential convergence for the following reason: as was said in the last section, the few-body ansatz explicitly breaks PH symmetry (although very slightly), with an amount proportional to the coupling terms between correlated and uncorrelated sectors that are not considered. As correlations become stronger, these terms become bigger and the symmetry breaking (that can be quantitatively seen in the occupation of the impurity level in the ground state deviating from $1/2$) is larger. In addition to this larger symmetry breaking, the emergent scale T_K is also exponentially smaller ($T_K \simeq 10^{-6}$ for $U = -1.0$), and the error suffice to polarize the system and cut correlations. Hence, a correlated space with at least $M = 8$ orbitals is required for the state to be correct. This effect is similar to the symmetry breaking in the NRG when $U \rightarrow U_c$ for too few kept states. For $U = -0.5$, this slight symmetry breaking allows more freedom for the construction of orbitals, and yield to slightly more accurate results.

In the right panel of Fig. 4.3, the spatial dispersion of the most correlated orbital $q_1 = \sum_i D_{i1} c_i$ (whose occupancy is the closest to $1/2$) is plotted in the correlated

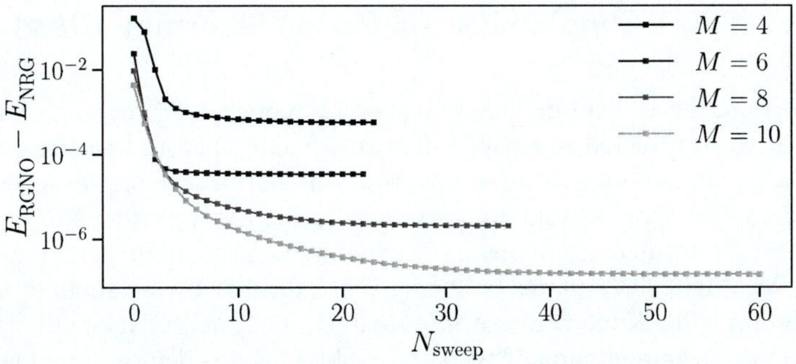

Fig. 4.4 Relative ground state energy (RGNO vs NRG) against the number of sweeps of the RGNO, for various numbers M of correlated orbitals. The computation is performed for $U = -0.5$, corresponding to $T_K \simeq 3 \times 10^{-3}$. The sweeping process is stopped when the difference of energy between two sweeps is less than 10^{-10}

regime $U = -0.5$ for various sizes of the correlated space, to be compared to the one computed with the NRG. Even for simulations with a number of correlated orbitals as small as $M = 6$, the orbital is nearly indistinguishable from the NRG result, and gets even more accurate when increasing M. The largest amplitudes are concentrated around the site $n = 15$, which corresponds on the Wilson chain to an energy of $\Lambda^{-15/2} = 5 \times 10^{-3}$, that corresponds roughly to $T_K \simeq 3 \times 10^{-3}$. As a side remark, we could think that it may be worth to diagonalize the Hamiltonian at the end of the flow in the enlarged space containing $M + 2$ orbitals instead of M. Actually, the algorithm tends to optimize the participation of orbitals in the correlated space with M orbitals, such that the occupancy of uncorrelated orbitals is the closest to 1 or 0. Hence, the most correlated orbitals of the uncorrelated sector have occupancies so close to 0 or 1 that including them in the exact diagonalization does not change the result.

Finally, it is shown in Fig. 4.4 the convergence of the energy at fixed M as a function of sweep iterations. The energy first decreases fast to the true value, then saturates to a plateau. Actually, the convergence criterion is arbitrary, since the sweeping process can be repeated indefinitely. For these simulations, the algorithm was stopped when the energy difference between two steps was around 10^{-10}, chosen to be far below T_K. This ensures the state to be converged, but too much effort may be put to converge a state at 10^{-10} when it actually converges to 10^{-4} above the true ground state energy. For larger M, more sweep are necessary since more degrees of freedom in the correlated sector need to be updated. In general, only a few tens of sweeps are required for convergence, and the procedure is even faster for less correlated states. Thus, the RGNO has proven to reproduce accurately the set of the M mostly correlated orbitals in the Wilson chain representation of the IRLM. However, it does not rely on any assumptions about the spectrum, so that it can be extended to real space lattices, thus exceeding the strongest limits of the NRG.

4.4 Real Space Simulations and Kondo Screening Cloud

In this section, the RGNO method is used to compute the screening cloud of the IRLM that was presented in Chap. 2. It was explained before that this observable requires a good resolution at all scales, from the lattice spacing up to the Kondo length L_K and beyond. Notable studies were presented, using either NRG or DMRG simulations: the former was numerically expensive, yet accurate at all scales, while the latter was limited to systems of hundreds of sites that are too small to obtain the full resolution of the Kondo crossover for realistic T_K. The RGNO method is a good candidate for the computation of such observables, as it is demonstrated now.

To study the screening cloud of the IRLM, the Wilson chain of the Hamiltonian (3.11) is changed to a real-space tight-binding chain with constant hopping $t_i = t$. For the density of states at zero energy to remain the same as in the box used for NRG simulations ($\rho_0 = D/2$), the hopping is chosen to be $t = D/2$. In this configuration, the screening cloud observable is still defined by the density-density correlator in Eq. (2.23). This study consider chains of $N = 10^4$ sites, such that energies of the order 10^{-4} are attainable. Without any simple benchmark with other methods on this observable at any interaction strength, the variance of the Hamiltonian is computed for a quantitative measure of the accuracy of the constructed few-body state. The variance is defined as:

$$\sigma = \frac{\langle H^2 \rangle - \langle H \rangle^2}{\langle H \rangle^2}, \tag{4.2}$$

where the averages are done with respect to the few-body wave function $|\Psi\rangle_{\text{few}}$. If Ψ is an exact eigenvector of the full Hamiltonian, *i.e.* $H|\Psi_n\rangle = E_n|\Psi_n\rangle$ and $H^2|\Psi_n\rangle = E_n^2|\Psi_n\rangle$, in such way that $\sigma = 0$. A well approximated state is expected to yield a small variance. In addition to be a check for the accuracy of the state that is being built, it can be a good criterion of convergence for the algorithm. However, the computation of the variance requires the exact diagonalization of terms involving 6 fermionic operators (details about the computation of the variance are given in appendix D) which is numerically expensive when $M \geq 10$. Moreover, the square root of the variance measures the error of the few-body energy against the real ground state, to be compared to the order of magnitude found in the benchmarking with the NRG. In Fig. 4.5, the convergence of the variance against the number of sweeps is computed for a real space chain of $N = 1000$ sites, for an interaction strength $U = -0.5$. The convergence is comparable to the one of the energy of Fig. 4.4, and it is seen that the algorithm converges faster than with the Wilson chain.

The correlator C_i is plotted for a chain of $N = 10^4$ sites in Fig. 4.6, for $U = -0.5$, $V = 0.15$ and $t = 0.5$, corresponding to a screening length $L_K \simeq 300$ sites. Even and odd sites have been separated in the bottom and top panels respectively to extract the envelope of the $2k_F$ oscillations of the cloud. Dash lines corresponding to the theoretical predictions [12, 13] for the short and long range scalings are also added. The crossover between the two regimes happens between $i = 100$ and $i = 400$ corresponding to the expected Kondo length. The computation of this observable is

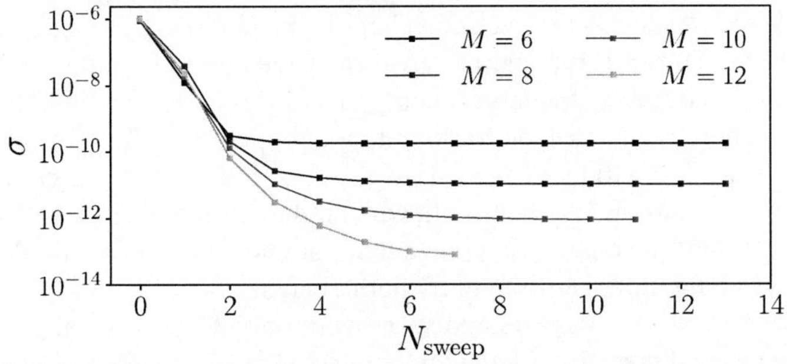

Fig. 4.5 Variance of the Hamiltonian at each sweep of the RGNO procedure for various M. The parameters for the simulation are $N = 10^3$ sites, $U = -0.5, t = 0.5$ and $V = 0.15$. The square root of the variance is proportional to the error of the variational energy to the true one of the ground state, that is seen to converge in less than 10 sweeps for any value of M

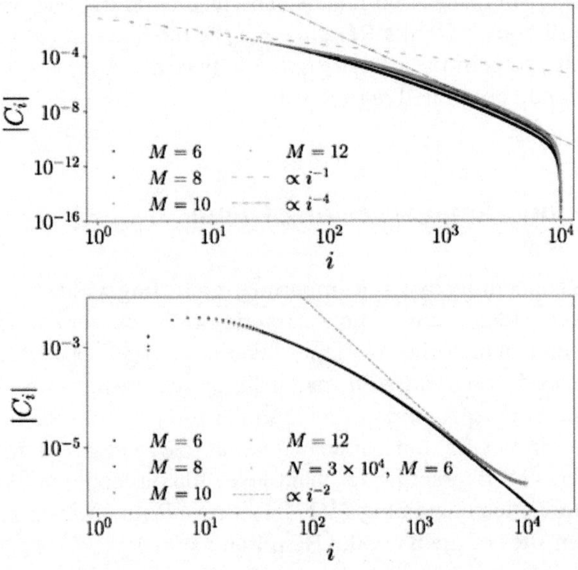

Fig. 4.6 Absolute value of the density-density correlator C_i computed at each site i, for a real-space tight-binding chain of $N = 10^4$ site, for increasing number of correlated orbitals M. The parameters of the IRLM Hamiltonian are $U = -0.5$ and $V = 0.15$, corresponding to $T_K \simeq 3 \times 10^{-3}$ (and $L_K \simeq 300$). Top and bottom panels separate the odd and even sites to reveal the envelopes of the oscillations of the screening cloud. The dashed lines are fitted power laws corresponding to the analytical scaling of the cloud at short and long range. In the bottom panel, a curve for a simulation of $N = 5 \times 10^4$ sites is shown, and fits better with the expected scaling. After $i = 2 \times 10^4$, the observable is not well converged and is cut for readability

only a single shot evaluation of each correlator in the few-body ground state obtained through the RGNO. While the scalings are clearly seen for the smallest component of the cloud, the i^{-2} decay of the largest component is less clear. Indeed, the crossover usually takes more than a decade for long range properties to fully develop, and the sum rule in Eq. (2.24) induces finite size effect in the tails of the cloud. A result for $N = 5 \times 10^4$ sites and $M = 6$ is shown and the scaling is clearer. However, the results at the end of the chain (not shown here) are not converged, simulations of so large systems being more sensible to numerical noise.

This curves show that a correlated space with only $M = 6$ correlated orbitals is enough to get a good precision for the screening cloud in a strongly correlated regime. For $N = 10^4$ sites, the one-body diagonalization required in the initialization take some time to be performed, and as $M \geq 8$ the simulations can take few-hours on a computer. However, for $M = 6$ and $N = 10^3$ sites, a converged state can be built in a few seconds. This is exploited in the next chapter for a statistical study of Kondo correlations in disordered metals. The results on the one-dimensional case were already well known from NRG [14], but the method can be used to study the two-dimensional problem on a large square lattice, that was not addressed in the literature in a Kondo correlated regime yet.

4.5 Two-Dimensional Screening Cloud

Usually, studies concerning dynamic impurities interacting with a two or three dimensional bath are considered under the assumption of a circular (or spherical) Fermi surface [15], within which the rotational invariance together with the short-range interaction reduces the problem to one radial dimension. In high dimensions, the large dimension of the Fock space prevented the microscopic study of Kondo screening clouds spreading in two or three dimensional lattices. The problem that is considered here concerns an interacting resonant level adatom coupled to the central site $(0, 0)$ of a square lattice containing $(2\Omega + 1)^2$ sites. With uniform nearest neighbor hopping between sites of the bath, the Hamiltonian reads:

$$\begin{aligned}\mathcal{H}_{2d} = U\left(d^\dagger d - \frac{1}{2}\right)\left(c^\dagger_{0,0}c_{0,0} - \frac{1}{2}\right) + V\left(c^\dagger_{0,0}d + d^\dagger c_{0,0}\right) \\ + \frac{t}{2}\sum_{\langle\langle x_1,x_2\rangle\rangle}\left(c^\dagger_{i_1,j_1}c_{i_2,j_2} + c^\dagger_{i_2,j_2}c_{i_1,j_1}\right),\end{aligned} \tag{4.3}$$

where $x_a = (i_a, j_a)$ is the set of coordinates of the sites of the lattice, and periodic boundary conditions are considered. The hopping term is normalized such that the half-band width is $2|t|$, as in the one-dimensional case. In this study, a lattice of 301×301 sites is considered, such that 90602 orbitals must be treated in the Hilbert space. However, the lattice enjoys point group symmetries with which some orbitals are redundant and can be generated by a smaller set of orbitals. The first symmetry are generated by reflections about the diagonal ($i = j$), such that $c_{i,j}$ generates $c_{j,i}$. The second corresponds to an invariance under $\pi/2$ rotations with respect to the central cite, which allows $c_{i,j}$ to generate $c_{-j,i}$, and by extension of the first symmetry $c_{-i,j}$.

An implementation of these symmetries would allow to divide the number of orbitals to be handled by 8. To reveal these simplifications, we start from the set of orbitals diagonalizing the Hamiltonian decoupled from the impurity ($V = 0$), which can be written:

$$\begin{cases} a_{m,n} = \frac{1}{\sqrt{2}} \sum\limits_{i,j=-\Omega}^{\Omega} \left(\psi_m(i)\psi_n(j) + \psi_m(i)\psi_n(j)\right) c_{i,j}, & \forall m \neq n \\ a_{m,m} = \frac{1}{\sqrt{2}} \sum\limits_{i,j=-\Omega}^{\Omega} \psi_m(i)\psi_m(j)\, c_{i,j}, \end{cases} \tag{4.4}$$

where elements of the expansion $\psi_n(i)$ are defined by cosine functions, as it is usual for non-interacting square lattices with nearest neighbor hopping:

$$\psi_n(i) = \frac{1}{\sqrt{\Omega+1}} \cos\left[\frac{\pi j(m+\frac{1}{2})}{\Omega+1}\right]. \tag{4.5}$$

The eigenenergy associated to the orbital $a_{m,n}$ is:

$$E_{m,n} = t\left(\cos\left[\frac{\pi(m+1/2)}{\Omega+1}\right] + \cos\left[\frac{\pi(n+1/2)}{\Omega+1}\right]\right). \tag{4.6}$$

At this point, symmetries are not accounted and the spectrum $E_{m,n}$ is degenerate. As the impurity only couples to the site (0, 0), and the Wannier orbital associated to this site $c_{0,0}$ is left invariant by the generators of both symmetries, we only want to define orbitals in the symmetric sector containing $c_{0,0}$. Let $E_{\alpha\in\{1,N\}}$ be an eigenenergy in the symmetric sector containing N orbitals, we can define the associated symmetric orbital as the sum of every eigenorbital $a_{m,n}$ with an energy $E_{m,n} = E_\alpha$, and we further want each orbital $a_{m,n}$ involved in this new orbital to be coupled to the impurity, i.e. $\langle 0|c_{0,0}a^\dagger_{m,n}|0\rangle \neq 0$. Such an orbital is easily defined as:

$$b_{E_\alpha} = \frac{1}{\sqrt{D_{E_\alpha}}} \sum_{m,n} \delta_{E_{m,n},E_\alpha} \langle 0|c_{0,0}a^\dagger_{m,n}|0\rangle a_{m,n}, \tag{4.7}$$

where $\langle 0|c_{0,0}a^\dagger_{m,n}|0\rangle$ can be explicitly expressed as $\sqrt{2-\delta_{m,n}}/(\Omega+1)$, such that the norm D_{E_α} can be written:

$$D_{E_\alpha} = \frac{\sum\limits_{m,n} \delta_{E_{m,n},E_\alpha}(2-\delta_{m,n})}{(\Omega+1)^2}. \tag{4.8}$$

Hence, we have a new reduced set of orbitals which are all left invariant by application of the symmetry and which couple to the impurity. Remarking that the Wannier orbital located at the site coupled to the impurity is the simply sum of these orbitals $c_{0,0} = \sum_{\alpha=1}^{N} \sqrt{D_{E_\alpha}} b_{E_\alpha}$, the Hamiltonian reads:

$$\mathcal{H}'_{2\mathrm{d}} = U\left(d^\dagger d - \frac{1}{2}\right)\left(\sum_{E_\alpha, E_\beta = E_1}^{E_N} \sqrt{D_{E_\alpha} D_{E_\beta}} b^\dagger_{E_\alpha} b_{E_\beta} - \frac{1}{2}\right) \\ + V \sum_{E_\alpha = E_1}^{E_N} \sqrt{D_{E_\alpha}} \left(b^\dagger_{E_\alpha} d + d^\dagger b_{E_\alpha}\right) + \sum_{E_\alpha = E_1}^{E_N} E_\alpha b^\dagger_{E_\alpha} b_{E_\alpha}. \tag{4.9}$$

Thus, the RGNO procedure is applied to the above problem concerning roughly 10^4 orbitals, as in the one dimensional study presented in the previous section. The correlated sector is built from the M b_E-orbitals with energies closest to 0 (by pairs around the Fermi energy). The difference lies in this initialization, where the single-particle Hamiltonian is diagonalized without the hybridization term V for good convergence. This may be due to a renormalization of the bare coupling V by the interaction, the interacting solution being thus closer to the Hamiltonian at $V = 0$.

The two-dimensional cloud $C_{i,j}$ is plotted in real space in the left panel of Fig. 4.7, for $V = 0.15$, $U = -0.4$ and $t = 0.5$, corresponding to a Kondo temperature of $T_\mathrm{K} \simeq 1.2 \times 10^{-2}$. In the right panel the same observable is plotted with the same parameters but at $U = 0$, corresponding to an enhanced Kondo temperature of $T_\mathrm{K} = 5 \times 10^{-2}$. Only the sites $0 \leq i, j \leq 100$ are plotted for readability, corresponding to $1/9^{\mathrm{th}}$ of the full simulated system. The correlated space is chosen to harbor $M = 12$ orbitals to reach an accuracy of order 10^{-9}, since correlations spreading in the bath are expected to be exponentially small. In both cases, the screening cloud appears mostly along the $(j, \pm j)$ diagonals, and decays exponentially as orbitals are far from the diagonal.

In the non-interacting case, correlations are always negative and are only non-zero in the A-sublattice for which $i + j$ is even. Hence, the charge polarity along the lattice is opposite to that of the impurity. This might indicate, for a spinful model, that only spins opposite to that of the impurity accumulate to offset it. In contrast, for the interacting case, the charge polarity is not opposite everywhere. There is a region spreading tens site away from the impurity exhibiting the same polarity as the impurity (in light grey in the figure), which is following the shape of the long ranged cloud on shorter scales. Another qualitative difference is that correlations in the B-sublattice are not zero anymore. A similar feature was found with the one dimensional screening cloud, whose odd sites correlations vanishes exactly at $U = 0$ only.

The main observation about this screening cloud is its strong anisotropy along the two directions of space. In the main diagonal, correlations extend further than the expected L_K if the rotational invariance is supposed, while it vanishes faster in the orthogonal direction. A similar anisotropic pattern around a static impurity

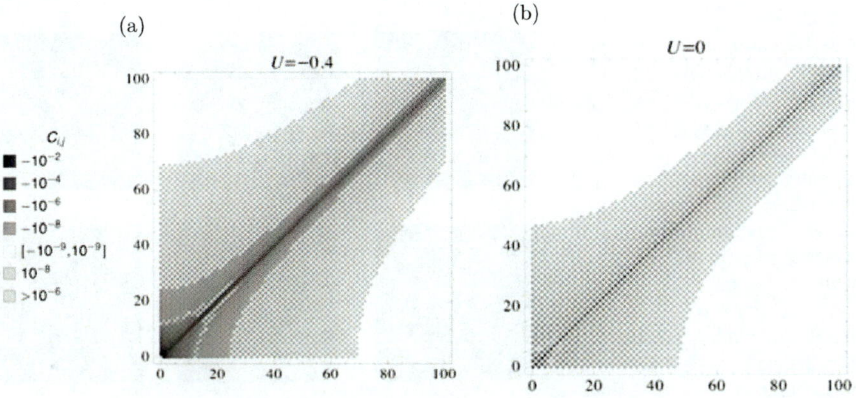

Fig. 4.7 **a** Two dimensional density-density correlator $C_{i,j}$ of the IRLM for $V = 0.15$, $U = -0.4$, $t = 0.5$ in a square lattice of 301×301 sites. **b** Same plot as **a** for $U = 0$, showing that correlations are always negative in the non-interacting case. Both screening clouds show the same kind of anisotropic spatial structure

embedded in a bath with a square Fermi surface as an effect of Friedel oscillations has been reported in Ref. [16]. From this finding, it can be supposed that on a half-filled square lattice with many IRLM-type impurities, some of them will interact through the conduction band at scales larger than L_K, while the majority will not even know if they are closer to each other. This may help to investigate anisotropic RKKY interactions, that is known to be closely linked to the screening process in the conduction band.

4.5.1 No Reduction to One-Dimensional Chains

The study presented above considered a fully-interacting problem on a $L \times L$ lattice, that was reduced to $L^2/8$ relevant orbitals taking advantage of point group symmetries of the square lattice. However, it was proposed in Ref. [17] that tight-binding Hamiltonian of higher dimensions maps onto one-dimensional chains of size $\mathcal{O}(L)$ with nearest neighbor hopping only. The proposed scheme recursively generates orthogonal Lanczos orbitals $|\Psi_n\rangle$, such that $\langle\Psi_m|H|\Psi_n\rangle = 0$ for $m < n - 1$. This scheme is exact once all symmetry class orbitals have been exhausted, (here $L^2/8$ orbitals) but it was claimed in Ref. [17] that only a linear amount of these orbitals L is enough to converge exactly the ground state. While the study of symmetries presented before ensures that the result can be exact only if $L^2/8$ orbitals are considered, it could be interesting to see if a certain amount of these orbitals are almost irrelevant, reducing the complexity to $\mathcal{O}(L)$ orbitals up to some numerical accuracy. The procedure presented in Ref. [17] has been implemented to verify this statement, for the IRLM at $U = 0$ on a two-dimensional diamond-shaped lattice, as was suggested in

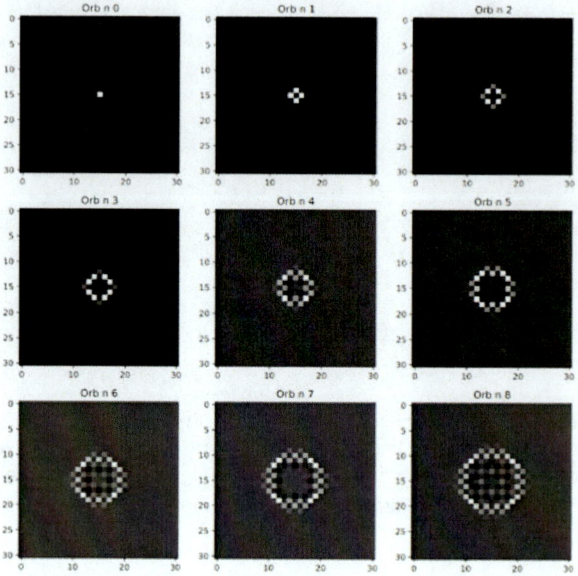

Fig. 4.8 Real space dispersion of the first 9 Lanczos orbitals obtained through the method of Ref. [17], for a lattice of 31 sites with a diamond-like geometry, with a uniform hopping $t = 0.5D$. Each Lanczos iteration generates an orbital $|\Psi_n\rangle$ in the symmetric sector, with $n < L^2/8$. For $n > L^2/8$, orbitals are combinations of the previous ones, and are thus redundant

the article. The parameters are $V = 0.15$, $t = 0.5$ and $L = 100$ sites, corresponding to 20201 sites in the lattice (without the impurity). Standard procedures of orthonormalization and normalization during the Lanczos scheme advised by Ref. [17] yield a stable algorithm for these parameters. In Fig. 4.8, the 9 first generated orbitals are shown, the first one being the seed of the recursive scheme.

To refute the statement of a chain of size $\sim L$ coupling to the impurity, we investigate the error in the hybridization $\langle d^\dagger c_{0,0}\rangle$ as the number of Lanczos orbitals increase compared to the full system (for a non interacting impurity), and the spatial dispersion of the hybridization between the impurity and the second site of the chain $|\langle d^\dagger c_{2j,0}\rangle|$ ($2j$ is used instead of j to avoid the cancellations of the correlator on the B-sublattice). In the left panel of Fig. 4.9, the precision after $L = 100$ iterations is only at 2 digits, and becomes correct to machine precision only when $L^2/8 \simeq 2600$ orbitals are generated. The result is confirmed by the right panel of Fig. 4.9: while short range correlations may be described by fewer orbitals, only the required amount of orbitals $L^2/8$ manages to converge the observable at all scales. A misunderstanding in Ref. [17] may be the following: for too small L ($L \sim 10^2$), the reduction due to symmetries $L^2/8$ is close to L. For 400 sites, the exact reduction accounts for 50 orbitals, while $L = 20$, that is of the same order of magnitude. However, it is clear that when L increases, the reduction to $\mathcal{O}(L)$ chains can not be reliable and no shortcut beyond point group symmetries can provide an accurate calculation of the two-dimensional screening cloud.

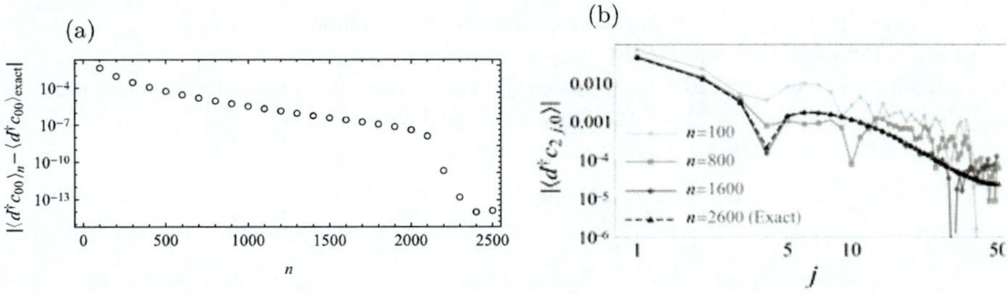

Fig. 4.9 **a** Difference between the local hybridization $\langle d^\dagger c_{0,0}\rangle$ computed with n Lanczos orbitals and the one computed with all eigenorbitals of the system, for the non-interacting resonant level on a diamond shaped lattice. Boundaries are at $|i| + |j| = 100$ such that they respect the underlying spatial symmetries. **b** Absolute value of the spatial dispersion of the hybridization along the j direction $\langle d^\dagger c_{2j,0}\rangle$ on the sublattice A only ($i + j$ even), for different number of Lanczos orbitals n (same parameters as panel **a**). This result clearly shows that accuracy is lost for a number of orbitals smaller than the one in the symmetric sector, and demonstrates that no efficient mapping to one-dimensional chain of size $\mathcal{O}(L)$ is possible

References

1. Bravyi S, Gosset D (2017) Complexity of quantum impurity problems. Commun Math Phys 356:451–500. ISSN: 1432-0916
2. Boutin S, Bauer B (2021) Quantum impurity models using superpositions of fermionic Gaussian states: practical methods and applications. arXiv:2105.01088 [condmat, physics:quant-ph]
3. Snyman I, Florens S (2021) Efficient impurity-bath trial states from superposed Slater determinants. arXiv:2106.14027 [cond-mat]
4. Zgid D, Gull E, Chan GK-L (2012) Truncated configuration interaction expansions as solvers for correlated quantum impurity models and dynamical mean-field theory. Phys Rev B 86:165128. ISSN: 1098-0121, 1550-235X
5. Lu Y, Höppner M, Gunnarsson O, Haverkort MW (2014) Efficient real-frequency solver for dynamical mean-field theory. Phys Rev B 90:085102. ISSN: 1098-0121, 1550-235X
6. Lu Y, Cao X, Hansmann P, Haverkort MW (2019) Natural-orbital impurity solver and projection approach for Green's functions. Phys Rev B 100:115134. ISSN: 2469-9950, 2469-9969
7. Go A, Millis AJ (2017) Adaptively truncated Hilbert space based impurity solver for dynamical mean-field theory. Phys Rev B 96:085139. ISSN: 2469-9950, 2469-9969
8. Kitatani M, Sakai S, Arita R (2021) Natural orbital impurity solver for real-frequency properties at finite temperature. arXiv:2107.06517 [cond-mat]
9. Li X, Paldus J (2005) Recursive generation of natural orbitals in a truncated orbital space. Int J Quant Chem 105:672–678. ISSN: 1097- 461X
10. He R-Q, Lu Z-Y (2014) Quantum renormalization groups based on natural orbitals. arXiv:1306.3662 [cond-mat, physics:physics, physics:quant-ph]
11. Yang C, Feiguin AE (2017) Unveiling the internal entanglement structure of the Kondo singlet. arXiv:1703.02383 [cond-mat]
12. Ishii H (1978) Spin correlation in dilute magnetic alloys. J Low Temp Phys 32:11 (1978)
13. Barzykin V, Affleck I (1998) Screening cloud in the k-channel Kondo model: perturbative and large-k results. Phys Rev B 57:432–448
14. Borda L (2007) Kondo screening cloud in a one-dimensional wire: numerical renormalization group study. Phys Rev B 75:041307. ISSN: 1098-0121, 1550-235X
15. Affleck I (2009) The Kondo screening cloud: what it is and how to observe it

16. Andrade EC, Miranda E, Dobrosavljevi ć V (2010) Quantum ripples in strongly correlated metals. Phys Rev Lett 104:236401. Accessed from 23 June 2010
17. Allerdt A, Feiguin AE (2019) A Numerically exact approach to quantum impurity problems in realistic lattice geometries. Front Phys 7. ISSN: 2296-424X

Part III
Quantum Impurities in a Disordered Environment

Chapter 5
RGNO Study of Screening Clouds in Disordered Environments

In Chaps. 3 and 4, natural orbitals of quantum impurity problems (both IRLM and SIAM) were investigated. They proved to efficiently describe the ground state of these systems in all regimes, provided that parameters are not tuned too close to a phase transition. Natural orbitals occupancies revealed a strongly hierarchical structure of correlations, whereupon an ansatz for the ground state wave function was proposed. A method dubbed RGNO (for Recursive Generation of Natural Orbitals) has been developed to generate the set of $M \ll N$ (N the number of sites in the system) mostly correlated natural orbitals, that are at the core of the wave function. This method is non-perturbative and allows to deal with systems in any one-particle basis, so that the development of the RGNO was motivated by real space simulations, that were expensive for usual NRG calculations. The efficiency of the algorithm was used to solve large systems with a high precision at every spatial scales. Computing spatial structures with well developed correlations in the non-interacting bath is relatively fast, so that accumulating statistics on fully many-body simulations in the thermodynamic limit becomes possible.

In Sect. 2.5, the literature concerning Kondo correlations in disordered environments was presented. The presence of charge disorder in the spinful Kondo model has been demonstrated to be responsible for non-Fermi liquid behaviors at low temperatures, where part of the local moments were not quenched by the surrounding electrons. Due to the difficulty of simulating a large amount of realistic lattices, approximate scaling equations have been mostly used in these studies. It was however enough to explain the non-Fermi liquid behavior for dilute Kondo alloys through the distribution of Kondo temperatures, yet these studies were limited to this only quantity.

In this chapter, the RGNO is employed to study the effect of charge disorder on Kondo correlations in the IRLM. The IRLM is chosen for convenience of implementation and because disorder in this model was not studied in the literature until now. We will also consider the disordered SIAM as an outlook in the end of this chapter. This study also serves as a testbed for the effectiveness of RGNO, which will prove to be suitable for this type of problem. Contrary to charge disorder in the

M. Debertolis, *Quantum Impurity Problems in the Framework of Natural Orbitals*, Springer Theses, https://doi.org/10.1007/978-3-031-47233-6_5

Kondo model, where correlations spread in further space, due to the depletion of the DOS, charge disorder in the IRLM is expected to shorten correlations. Natural orbitals are then expected to be more localized around Wannier ones, which would help the convergence of the algorithm. The distribution of Kondo temperatures will be presented, in addition to relevant non-local observables that will help to better understand the physics involved.

5.1 Statistics on Disordered Quantum Impurities

The charge disorder is added to the real space IRLM with uniform hopping along the chain $t = D/2$ as a random onsite potential v_i. The Hamiltonian reads:

$$\mathcal{H}_{\text{dis}} = \mathcal{H}_{\text{IRLM}} + \sum_{i=0}^{N-1} v_i \, c_i^\dagger c_i \,, \tag{5.1}$$

and no potential is added on the impurity level $i = -1$ that would polarize it trivially. On each site of the chain, the local potential v_i is uniformly distributed in the interval $[-v, v]$. These terms explicitly break PH symmetry, and the average onsite charge $\langle c_i^\dagger c_i \rangle$ deviates randomly from $1/2$. The weak symmetry breaking by the ansatz is thus irrelevant in this context and does not affect the results. This charge potential are equivalent to random uniaxial magnetic disorder in the Kondo model that has not been studied yet, as it is not as common as charge disorder that is inherent to experimental samples.

In the following, the parameters of the Hamiltonian for each disorder realization are chosen to be: $U = -0.5$, $t = 0.5$, $V = 0.15$, $M = 6$ and $N = 10^3$ sites. Various disorder strengths v will be considered to explore the effect of stronger (or weaker) perturbations. As correlations are expected to become shorter with the presence of disorder, $N = 10^3$ sites are enough for correlations to develop, the clean Kondo scale being $L_{\text{K}} \simeq 3 \times 10^2$. The statistics are performed (for each set of parameters) for 10^4 realizations of the disorder, which provides smooth distribution functions.

The convergence of a state with $M = 6$ correlated orbitals has been checked by repeated simulations with increasing M for random realizations of the disorder, and $M = 6$ has proven to give well converged screening clouds and T_{K} (the usual convergence criterions are the ground state energy and the variance) in every case. For each run, the flow of the ground state energy and of a relevant observable (in this case the expectation value $\langle d^\dagger d \rangle$ through which T_{K} is computed) is followed and sweeps are stopped when both observables saturate. The variance saturates similarly to the ground state energy and will not be computed for evident computational cost reasons (see the section concerning the variance in Appendix D).

While most of the disorder realizations were easily converged by the RGNO (between a few sweeps and few tens of sweeps), some rare realizations (some tens out of the 10^4 realizations, corresponding to few permille) led to convergence issues after more than hundreds of sweeps. These problems of convergence arise with disorder when the energetic landscape is not smooth anymore, and local minima can be accidentally close to the true global ground state. Hence, the iterative process can be stuck in one of these minima and give a wrong result. This effect is clearly seen in some observables: for instance, it happened that $\langle d^\dagger d \rangle$ converged to some value for the first tens of sweeps, then jumped to another one, went back to the first value etc. This indicates that the picture of local minima is at play. To circumvent this issue, a solution was found: if a potential ϵ_d is added at the beginning of the flow (positive and negative) and quenched along the sweeps (or through several restarts of the algorithm), the algorithm starts with a better guess and manage to find its way (both signs are explored, and the ground state energy is compared to see which solution is better) to the actual ground state. However, as problematic runs can not be guessed a priori (no special pattern in the disorder has been isolated to be particularly difficult), the procedure should be applied to each run and would slow down non-problematic simulations. As these rare realizations were not representative of a specific part of the results (neither only large or small $T_{\rm K}$ for instance) they were discarded in the final results.

5.2 Distribution of Kondo Temperatures

The fate of Kondo temperatures is firstly examinated as it is the principal scale in clean systems. The distribution of $T_{\rm K}$ is not supposed to be Gaussian, the distribution is therefore studied as a whole instead of the first few moments. The results found in previous studies presented in Chap. 2 concerning the disordered Kondo model are briefly recalled here. When the disorder strength is small, the distribution follows a log-normal law, centered around the clean $T_{\rm K}$ with logarithmic tails at lower $T_{\rm K}$. As the disorder becomes stronger, the tails become bigger and bigger and a larger fraction of local moments is unscreened. The mechanism behind these remanent magnetic moment is the disorder-induced depletion of the density of states of the bath around the Fermi level, which prevents any state to couple to the impurity (whose chemical potential is at resonnance with the Fermi level of the bath) and thus to compensate the local spin.

For the disordered IRLM, the distribution of Kondo temperatures for various disorder strengths v is plotted in Fig. 5.1, and the phenomenon appears to be different from the spinful case. Kondo temperatures are still computed as the inverse of the local susceptibility of the impurity level (see Eq. (2.20)), such that two RGNO runs are necessary for each disorder realization. As expected when the disorder is strong enough, correlations spread less far than in the clean system, and Kondo temperatures drastically increase compared to the clean $T_{\rm K}^0$. Logarithmic tails are also reported for $T_{\rm K} \to T_{\rm K}^0$ as in the spinful problem, that seems inherent to Kondo systems. For smaller disorder strengths (but finite, $v \to 0$ becomes trivial again and the effect of disorder is not relevant), the distribution is bimodal: most of the realizations see their

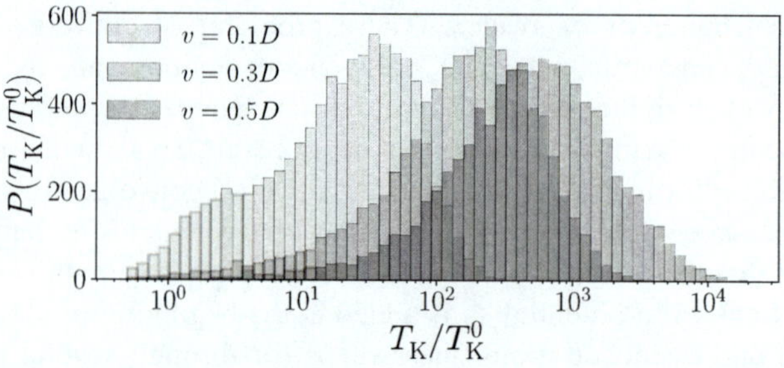

Fig. 5.1 Distribution of Kondo temperatures of the disordered IRLM for $U = -0.5$, $V = 0.15$, $N = 10^3$ sites and three different disorder strengths v. Kondo temperatures are normalized by the clean value $T_K^0 \simeq 3 \times 10^{-3}$. At strong disorder, the distribution is peaked at high values of T_K with logarithmic tails to T_K^0. A bimodal distribution is observed at weaker disorder, showing the robustness to disorder of a part of realizations

T_K increase as expected, but another significant part of the realizations keep a T_K close to the clean T_K^0.

For the runs with $T_K \gg T_K^0$, the mechanism is the following: local charges along the chain are driven away from their clean expectation value $\langle c_i^\dagger c_i \rangle = 1/2$. Thus, the impurity is polarized by a strong Hartree shift coming from its Coulomb coupling with the first site of the chain $U(\langle c_0^\dagger c_0 \rangle - 1/2)d^\dagger d \neq 0$. Insights can be drawn from the non-interacting problem: in Fig. 2.1, the susceptibility of the impurity level (that is the slope of the curve of the occupancy) is smaller when the impurity is almost full or empty ($n_d \simeq 1, 0$) which yields higher T_K. In addition, disorder induces random fluctuations in the density of states that may become highly asymmetric, which also polarizes the impurity through Eq. (2.14). Although these explanations are the most favorable to happen and clarify the part of $T_K \gg T_K^0$, the picture is missing for the remaining $T_K \simeq T_K^0$. We may argue that accidental realizations have only a small potential shift on the first site, and that the density of states remains mostly symmetric, but the question concerns more the *quantity* of realizations close to the clean case. To learn more from the effect of disorder on this system, the individual screening clouds are studied now.

5.3 Effect of Disorder on Spatial Correlations

5.3.1 Screening Clouds

Each screening cloud is computed through the correlator (2.23) as in the clean case. Showing every sample would be impossible, and the averaged screening cloud would be meaningless, so that three typical screening clouds are plotted in Fig. 5.2 to be compared to the clean one. These clouds are representative respectively of runs with large Kondo temperatures $T_K \gg T_K^0$, of almost unchanged $T_K \simeq T_K^0$ and of an

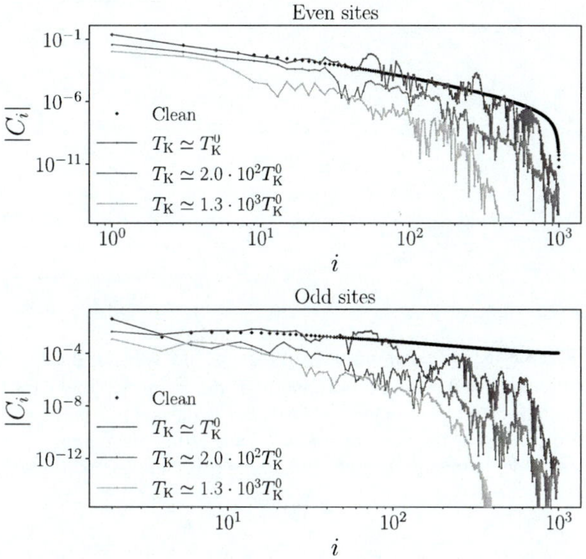

Fig. 5.2 Spatial correlator $|C_i|$ for $i \geq 0$ for the same parameters as Fig. 5.1 with $v = 0.3$. Top and bottom panel shows separately even and odd sites respectively for readability. Three different realizations corresponding to different values of T_K are shown, to be compared to the clean correlator represented by black crosses. The even/odd oscillations are not surviving as in the clean case, but the shape of the cloud (and the decay length of correlations) is not as much affected as the corresponding T_K

intermediate situation between the two, for $v = 0.3D$. From the 100 and 1000 fold enhancement of T_K between the different disorder realizations, these three clouds are expected to vanish clearly at different spatial scales, following the relations used in clean systems $T_K \simeq 1/L_K$.

However, what is seen in the figure is totally different: each observable has a similar shape and its spatial extent seems not radically affected by disorder despite the large local fluctuations. It was verified for more than these three samples that regardless of the T_K, the correlation length L_K is not reduced in proportion to the inverse of the enhancement of T_K. The definition of the Kondo temperature related to the local susceptibility $T_K = 1/4\chi$ is hence no more related to the spatial extent of correlations into the bath, and only to the local dynamics of the impurity level.

5.3.2 *Cloud Amplitude*

The effect of disorder is in fact stronger on the amplitude of the cloud. At a same site i, the different realizations give different orders of magnitude for the cloud $|C_i|$: a quantitative measure of this decrease is the summed total amplitude of the cloud without the impurity level, that would otherwise always return 0 due to the sum rule (2.24). As usual, the Toulouse point $U = 0$ is firstly investigated to build intuition (it is recalled that $T_K \neq 0$ in the non-interacting case). In this case, the ground state is a single Slater determinant such that Wick's theorem can be used to simplify Eq. (2.23):

$$C_{i \geq 0} = -|\langle c_i^\dagger c_i \rangle| < 0, \tag{5.2}$$

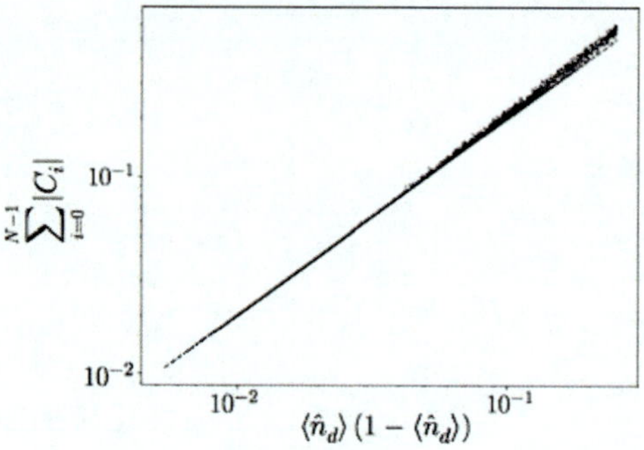

Fig. 5.3 Summed amplitudes of the absolute value of the correlator C_i deprived from the impurity site, against the folded occupancy of the impurity. The sum is sampled for the 10^4 realizations with the same parameters as Fig. 5.2. The dashed line corresponds to $2n_d(1-n_d)$, that is shown in Eq. (5.3) to be the exact result for $U=0$. While the match is perfect when the impurity is polarized, small deviations to the non-interacting result appear as $n_d \to 1/2$

while $C_{-1} = \langle \hat{n}_{-1} \rangle - \langle \hat{n}_{-1} \rangle^2 > 0$ (with $\hat{n}_{-1} = d^\dagger d$ the occupancy operator such that $\langle n_{-1} \rangle = n_d$). The last equation is valid at any U since $0 < \langle \hat{n}_{-1} \rangle < 1$, and is equal to $1/4$ at PH symmetry, and to 0 if the impurity is completely polarized. The sum rule (2.24) is used to derive the total cloud amplitude in the chain:

$$\sum_{i=0}^{N-1} |C_i| = 2\left(\langle \hat{n}_{-1} \rangle - \langle \hat{n}_{-1} \rangle^2\right) = 2n_d\,(1 - n_d). \tag{5.3}$$

The cloud amplitude is then totally controlled (for $U = 0$) by local properties on the impurity, that is directly linked to the Kondo temperature as we define it. Equation (5.3) is exact if every correlator $C_{i \geq 0}$ remains negative even for $U \neq 0$. In Fig. 5.3, this sum is computed at finite U for every realization of the disorder against the corresponding occupancy of the d-level to see how it deviates from the exact relation Eq. (5.3). For most of the samples, the statement stays true, and as n_d goes to $1/2$ $\left(n_d\,(1 - n_d) \to 1/4\right)$ becomes less precise. These realizations, corresponding to $T_\text{K} \simeq T_\text{K}^0$ have indeed several positive correlators (aligned with the impurity), but the deviation stays small. Therefore, for all realizations of disorder, the global cloud amplitude is controlled (or mostly controlled) by the local occupancy of the dot, which may be counter-intuitive at first sight. As U is getting exponentially closer to U_c, more and more correlators are expected to be aligned with the impurity, such that the picture of the occupancy controlling the cloud breaks down at $U = U_c + 0^+$, but it remains correct for more physical regimes as the one studied here.

5.3.3 Spatial Dispersion of Natural Orbitals

Correlations can also be investigated through the spatial profile of NOs. In the clean case, it was shown in Fig. 3.8 that the spatial dispersion of the most correlated NOs are highly sensible to correlations in the system. The amplitude of the orthogonal matrix defining the most correlated NO D_{i1} (and similarly its PH conjugate D_{i-1}) resolve the relevant scales by an enhancement of its weight around $i \sim L_K$. For the disordered problem, the spatial dispersion of the most correlated NO $|D_{i1}|$ is plotted in Fig. 5.4 for two realizations of disorder, one with $T_K \simeq T_K^0$ and another with $T_K \gg T_K^0$. The absolute value is plotted to avoid irrelevant sign changes, and is preferred to the square that crushes the lowest values to zero, for readability. In the first case (unchanged T_K), correlations survive almost until $L_K \sim 1/T_K$, but strong fluctuations are observed compared to the clean curve. In the opposite situation (increased T_K), the second curve has all its weight concentrated on the first ten or twenty sites, but this is still a larger range than $1/T_K$, that corresponds to the first few sites only.

To conclude this study concerning spatial correlations, local properties (n_d, T_K) and global ones (the screening cloud) were shown to be not as closely related as in the clean system, but remanent effects are however seen in some observables (spatial amplitude of the most correlated NO D_{i1}, total amplitude of the cloud). Hence, the single scale T_K is insufficient to provide an exhaustive picture of Kondo correlation in dirty samples, and the support of additional observables are required for this purpose. The observables presented in this section confirm the existence of rare realizations in which correlations are partially robust to disorder, and that the relevant quantity controlling this effect is the occupancy of the dot n_d. This quantity will be now investigated in order to understand more deeply the bimodal structure of the distribution of Kondo temperatures.

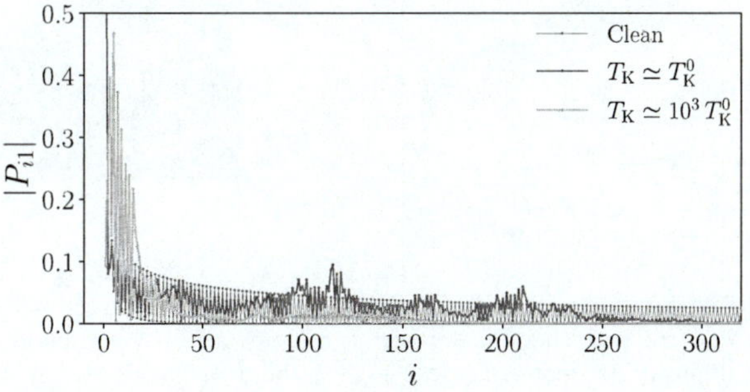

Fig. 5.4 Spatial dispersion of the most correlated NO for the same parameters as Fig. 5.2. The results for two different realizations of the disorder leading to a small and a large T_K are plotted, and the additional black curve corresponds to the NO in the clean system. The spatial extent of the NO with a large T_K is much shorter than the other: compared to the screening clouds, the spatial extent of NOs are more sensitive to the spatial scale defined by $1/T_K$

5.4 Local Charge Distribution

Similarly to screening cloud or the Kondo temperature, the average charge on the impurity level $n_d = \langle \hat{n}_d \rangle$ is extracted from each simulation. In the last section, it was shown that the charge of the impurity was closely related to the global amplitude of the correlation cloud. It was also shown that the bigger the global amplitude of the cloud, the lower the effective T_K. In the top left panel of Fig. 5.5, the Kondo temperature of each system is plotted against the corresponding occupancy of the impurity n_d. The result is as expected: the two quantities are correlated, and the closer is n_d to 1/2, the lower is the T_K. For n_d close to 0 and 1, the T_K are larger since the impurity is polarized. The Kondo temperatures are plotted against the folded occupancy $n_d(1 - n_d)$ are plotted in the bottom left panel of this figure: the information of $n_d \simeq 0, 1$ is thus concentrated around 0. In the top right panel of the

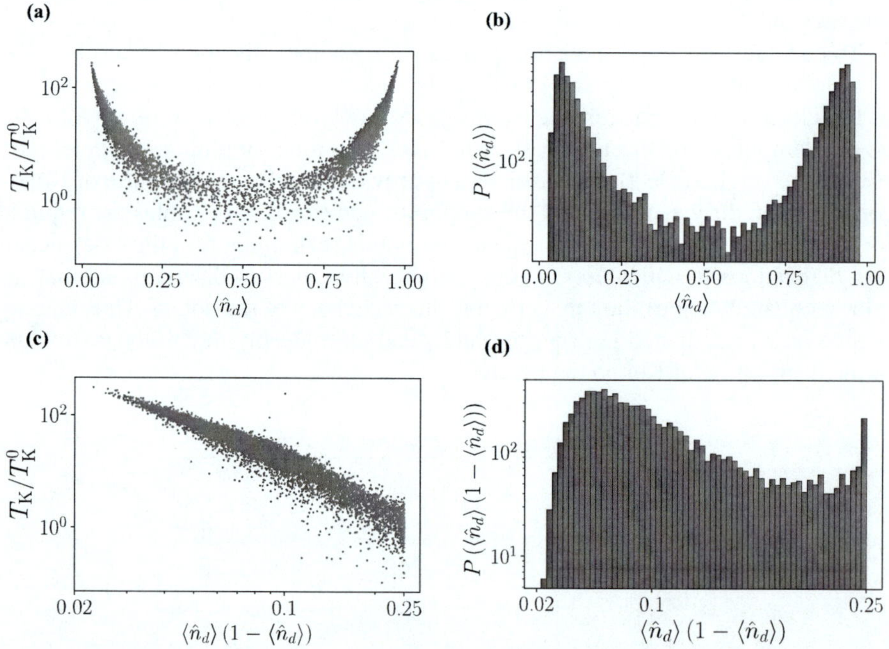

Fig. 5.5 **a** Kondo temperature against the impurity occupancy n_d for every 10^4 samples. The two variables are clearly correlated when the impurity is more polarized, corresponding to large values of T_K. When the impurity occupancy is close to 1/2, the dispersion is wider, but it is clear that these realizations correspond to those holding a low T_K. **b** Distribution of the occupancy of the d-level in the same conditions as (**a**), in log scale. A plateau is observed around $n_d \simeq 1/2$, which is responsible for the bimodal distribution of Kondo temperatures. **c** Same plot as (**a**) yet with the folded occupancy $n_d(1 - n_d)$ of each disorder realization, which shows the correlation between the two observables clearer. **d** Distribution of folded occupancies. A peak around 1/4 appears due to the folding of n_d. It is hence expected that Kondo temperatures are functions of the folded occupancy, the values around 0 and 1 corresponding to the same information about correlations

same figure, the distribution of the occupancy of the impurity is plotted. A plateau is observed around $n_d \simeq 1/2$ (although $n_d = 1/2$ never happens, and corresponds to the strict PH symmetry), and two peaks around 0 and 1. As for the previous plot, the distribution of the folded occupancies is plotted in the bottom right panel, in log-log scale for a good resolution around 0. Indeed, the impurity folded occupancy goes exponentially close to 0 (up to 10^{-2}) with a bell shaped curve centered around $\simeq 6 \times 10^{-2}$. The second peak seen around 1/4 appears only because the variable has been folded and the graph is in log-scale, the bins being larger around 1/4.

Nevertheless, even if the peak in the folded occupancy distribution is artificial, the plateau in the unfolded histogram is the one responsible for the bimodal structure of the distribution of Kondo temperatures. Indeed, the dispersion shown in the left panel has a finite width, and small T_K appears for $0.25 \lesssim n_d \lesssim 0.75$, so that the accumulation of all realizations in this interval participate to low T_K almost similarly. The naive picture would have been a peak around 0 (two peaks around 0 and 1 in the unfolded picture), and tails vanishing to 1/4 (to 1/2) without the plateau, that would not lead to a bimodal structure of T_K. The effect of the occupancy of the impurity on long ranged correlations is even stronger in the non-interacting problem where Eq. (5.3) is always exact. The free disordered problem still contains Kondo correlations (it corresponds to the Toulouse point of the Kondo model at finite J), and will be investigated in the following section.

5.5 Localization Length: One-Body Perspective

The effect of disorder on correlation properties is usually associated to strong localization effects. It was firstly investigated by Anderson in the seminal paper concerning the now called Anderson localization [1], which predicted an exponential localization of wave functions preventing transport for certain conditions of the disorder. The diffusion processes can occur in different channel of transport, and each of these are associated to a localization length, which corresponds to the spatial scale at which wave functions of electrons can spread as the system evolves ($t \to \infty$). These properties are associated to non-interacting systems with short range hopping and have been extensively studied in the past decades, and the reader can refer to the following references for a review [2–5].

For the scope of this study, only the localization length of the system for each realization of the disorder is looked for. Indeed, the results obtained above may be explained by localization properties of electrons inside the reservoir, that could take the lead over the Kondo effect. To compute the localization length of the interacting system, the scattering matrix must be known. However, computing this matrix in NRG with the disorder is a tough task, and for the RGNO an implementation of dynamical properties is required. Hence, the Toulouse point $U = 0$ will be investigated in this case, and the corresponding results will be confronted to the ones in the interacting and non-interacting cases, which also holds Kondo correlations.

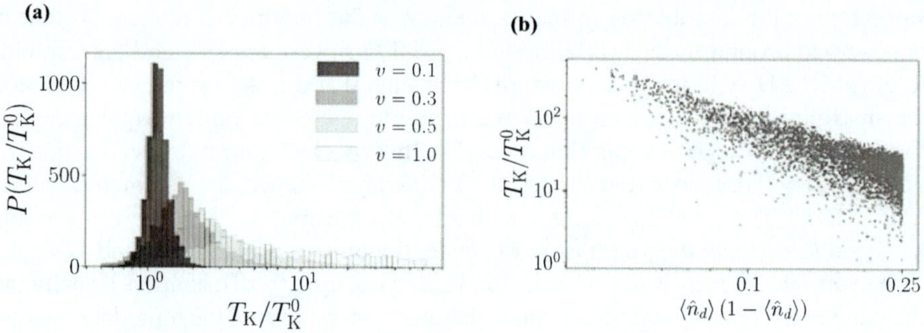

Fig. 5.6 **a** Distribution of Kondo temperatures for the disordered non-interacting RLM for four disorder strengths v, with $V = 0.15$, $t = 0.5$ and $N = 10^3$ sites. With these parameters, the clean Kondo length is $T_K^0 = 3 \times 10^{-2}$, which is 10 times larger than for the interacting problem at $U = -0.5$. **b** Kondo temperatures for each disorder realization against the corresponding folded occupancy of the impurity n_d

As a start, the principal results of the study in the interacting problem are reproduced for the non-interacting case. In the left panel of Fig. 5.6, the distribution of Kondo temperatures is plotted for four disorder strengths v. The main differences compared to the interacting case are the stronger disorder required to polarize the impurity, and the larger amount of impurities showing a $T_K \simeq T_K^0$. Indeed, the Kondo temperature in the clean free case is $T_K \simeq 3 \times 10^{-2}$, which is 10 times larger than for $U = -0.5$. This is inversely proportional to the susceptibility of the local charge to external fields, which makes it more robust to perturbations. Apart from these quantitative discrepancies, the behavior in the clean case is qualitatively similar to the one in interaction: the impurity is more and more polarized as the disorder increases, and less and less realizations keep a small T_K. However, bimodality is not seen at $U = 0$ for these strengths of disorder. In the right panel of Fig. 5.6, it is confirmed that the occupation of the impurity is directly related to Kondo temperatures.

The effect of disorder-induced localization is now investigated through the localization length, which can be obtained with the help of transfer matrices in the case of quadratic Hamiltonians. Transfer matrices are used to describe the scattering processes across multiple impurities, that will be modeled in our problem by the random onsite potential along the tight-binding chain that contains only one channel. The discretized stationnary Schrödinger equation can be written in the basis of Wannier orbitals $|n\rangle$ as:

$$\left(V\left(|-1\rangle\langle 0| + |0\rangle\langle -1|\right) + \sum_{n=0}^{N-1} v_n |n\rangle\langle n| + t\left(|n\rangle\langle n+1| + |n+1\rangle\langle n|\right)\right)|\Psi\rangle = E|\Psi\rangle. \tag{5.4}$$

An eigenvector of the Hamiltonian can be decomposed on Wannier orbitals as $|\Psi\rangle = \sum_n \Psi_n |n\rangle$, such that projecting Eq. (5.4) on any site $|n\rangle$ yield the following recursion relation:

$$\Psi_{n+1} = \frac{E - v_n}{t} \Psi_n - \Psi_{n-1}, \tag{5.5}$$

and by inserting the trivial identity $\Psi_n = \Psi_n$, the recursion relation can be written in a vectorial form:

$$\begin{pmatrix} \Psi_{n+1} \\ \Psi_n \end{pmatrix} = \begin{pmatrix} \frac{E-v_n}{t} & -1 \\ 1 & 0 \end{pmatrix} \begin{pmatrix} \Psi_n \\ \Psi_{n-1} \end{pmatrix} = \mathbf{T}_n \begin{pmatrix} \Psi_n \\ \Psi_{n-1} \end{pmatrix}. \tag{5.6}$$

The transfer matrix $\mathbf{T}_n$ has been introduced to link an element of an eigenvector to the next one. The total scattering matrix, that is related to the global localization properties of the system, is defined as the product of each transfer matrix $\mathbf{S} = \prod_{n=-1}^{N-1} \mathbf{T}_n$. This quantity can be scaled by a characteristic lengthscale appearing through the exponential of $\ln(\mathbf{S})$ (details can be found in Ref. [5]):

$$\mathbf{S} = e^{-Na/\xi_{\text{loc}}}, \tag{5.7}$$

with $a = 1$ the lattice spacing, and the introduced lengthscale corresponds to the correlation length defined as $\xi_{\text{loc}} = -N/\sum_{n=-1}^{N-1} \ln(T_n)$. A similar expression can be derived less phenomenologically with the Lyapunov exponents of the system (linked to the matrix S) [6]. Each transfer matrix T_n is simplectic, such that their product S stays simplectic, which imposes for eigenvalues of S to come in pair (s, s^{-1}). Hence, Eq. (5.7) reveals one ξ_{loc} and its inverse ξ_{loc}^{-1}, corresponding to short and long range decay length. In the following, the scale ξ_{loc} that interests us is defined through the minimum eigenvalue s_{min} of S as $\xi_{\text{loc}} = |N/\ln(s_{\text{min}})|$.

In the left and right panels of Fig. 5.7, the Kondo temperatures of the interacting and non-interacting problems are respectively plotted against the localization length, calculated for the corresponding realizations in the free problem for both cases. In both cases, the disorder strength is $v = 0.3$, such that the localization length is comparable to the clean Kondo scale L_K. What is however interesting besides the

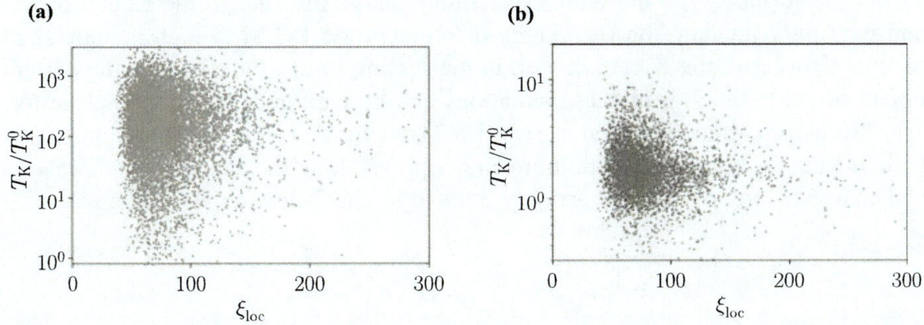

Fig. 5.7 **a** Kondo temperatures of the interacting problem ($U = -0.5$) against the localization length computed for the free problem, with the same disorder realization. **b** Same figure as (**a**) but for the Kondo temperatures of the free problem

absolute value of ξ_{loc} is that for both cases, ξ_{loc} is not correlated to the corresponding Kondo temperature T_{K}. The comparison at $U = -0.5$ is delicate, since the interaction may change the localization properties of the free problem. Nevertheless, the results at $U = 0$ confirm that localization effects of the wave function can not explain the delocalization of correlations in the system, if the disorder is not infinitely strong.

To summarize, the presence of disorder in the non-interacting bath of electrons renormalizes the Kondo temperature of the corresponding impurity, but correlations between the localized moment and the bath preserve their spatial structure, with however a decreasing amplitude as the disorder increases. The distribution of Kondo temperatures is bimodal for intermediate values of v, which has been related to the folded occupation of the impurity level $n_d(1 - n_d)$, whose distribution shows the same bimodality. The same results holds for the non-interacting problem, that is the Toulouse point of the Kondo problem for which $J > 0$. In the subsequent chapter, a simpler model of random matrices is analyzed and both numerical and analytical results are presented in different regimes.

5.6 Charge and Spin Disorder in the SIAM

To end this chapter, we present preliminary results concerning correlations around a spinful impurity surrounded by a disordered bath, modeled by the SIAM with either charge or spin disorder. The SIAM has been simulated with the RGNO in the same fashion as the IRLM, with the additional conservation of the total spin. With such an implementation, the orbitals of the uncorrelated sector are added four by four into the correlated space, corresponding to the spin up and spin down pair of PH-conjugated orbitals. The number of orbitals in the correlated sector is also required to be a multiple of four for the symmetries. The convergence is similar to the one of the IRLM, but with enlarged correlated sectors as spin comes into play. However, the additional spin symmetry reduces the amount of available states in a given sector, such that correlated spaces containing as much as 16 or 20 orbitals can be treated.

In order to unify the literature concerning charge disorder in the Kondo model and our study focusing on the charge disorder in the IRLM (which is equivalent to spin disorder in the Kondo model, in the scaling limit), the SIAM is the perfect object of study as both of these situations can be explored, when the Schrieffer-Wolff transformation presented in Sect. 1.3.3 is valid. The clean Hamiltonian $\mathcal{H}_{\text{SIAM}}$ is defined as in Eq. (3.11), with uniform hoppings along the chain $t_i = t$. The spin and charge disordered models are then defined by the following Hamiltonians:

$$\begin{aligned}
\mathcal{H}_\sigma &= \mathcal{H}_{\text{SIAM}} + \sum_{i=0}^{N-1} v_i \left(c_{i\uparrow}^\dagger c_{i\uparrow} - c_{i\downarrow}^\dagger c_{i\downarrow} \right), \\
\mathcal{H}_\rho &= \mathcal{H}_{\text{SIAM}} + \sum_{i=0}^{N-1} v_i \left(c_{i\uparrow}^\dagger c_{i\uparrow} + c_{i\downarrow}^\dagger c_{i\downarrow} \right),
\end{aligned} \tag{5.8}$$

with v_i defined as in Eq. (5.1). The parameters chosen for this study are $V = \sqrt{2D\Gamma/\pi}$, with $\Gamma = 10^{-1}D$, $t = D/2$ and $U = 9\Gamma$, in such a way that the Kondo temperature of the clean system is $T_K^0 \simeq 3 \times 10^{-3}D$, which ensures the scaling limit to be reached and the Kondo length to fit in the system ($N = 10^3$ sites). As before, every energy is written in units of the half bandwidth $D = 1$ that will be omitted in the following. For each realization of the disorder, the effective Kondo temperature is computed as the inverse of the local spin susceptibility:

$$T_K = \frac{1}{4\chi}, \quad \text{with} \quad \chi = \left.\frac{\partial \langle s_z \rangle}{\partial B_z}\right|_{B_z=0}, \tag{5.9}$$

where $s_z = n_{d\uparrow} - n_{d\downarrow}$ is the local impurity moment and B_z is a magnetic field on the impurity, which acts as $B_z s_z$ in the Hamiltonian. In Fig. 5.8, the distribution of Kondo temperatures for spin (left panel) and charge (right panel) disorder are plotted, for a strong disorder $v = 0.5$, as in the study concerning the IRLM. With the chosen parameters, the localization length is expected to be of the order of the Kondo length, so that strong effects on correlations are expected.

In the left panel, corresponding to the SIAM with spin disorder, the distribution of Kondo temperatures exhibits a bimodal structure similar to the one observed in the IRLM. This result is expected as these two models not only share the same non-interacting limit, within which the bimodal structure is still present, but are also equivalent in the strongly correlated regime ($U \gg \Gamma$ for the SIAM and $U \geq U_c$ for the IRLM). The first peak at $T_K \simeq 10\,T_K^0$ corresponds to the part of impurities whose correlations are only slightly affected by the disorder, and the second peak at $T_K \simeq 10^2\,T_K^0$ corresponds to charge-polarized impurities, whose dynamics are almost frozen. In the right panel, the charge-disordered case is plotted for the same

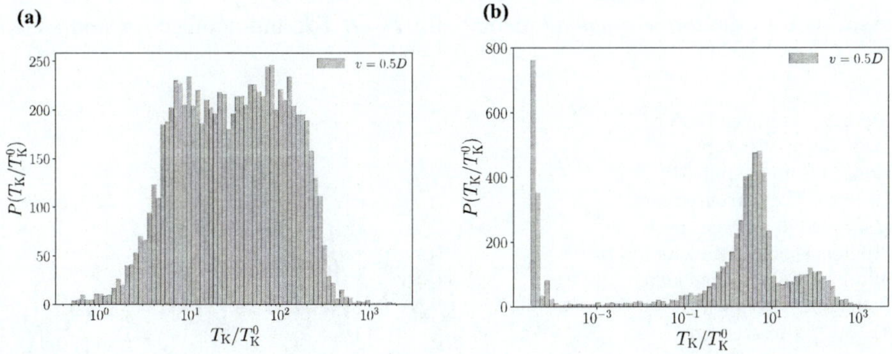

Fig. 5.8 **a** Distribution of the Kondo temperatures of the SIAM with spin disorder $\mathcal{H}_\sigma$, for $v = 0.5D$, $N = 10^3$ sites and 10^4 realizations of the disorder. This distribution is similar to ones of the disordered IRLM presented in Fig. 5.1. **b** Distribution of the Kondo temperatures of the SIAM with charge disorder $\mathcal{H}_\rho$ for the same parameters. This distribution is to compared to the one in Fig. 2.7, calculated in Ref. [7]. The trimodal structure was not predicted in previous studies

large disorder $v = D/2$. In this setup, a trimodal distribution is observed, with the same two peaks as in the spin-disordered case, and an additional sharper peak for $T_K \simeq 10^{-4}\,T_K^0$.

In Refs. [7–10], the study of the Kondo model did not allow charge fluctuations (n_d is fixed to one), such that charge-frozen configurations with $T_K \gg T_K^0$ are not captured in theses studies, but they observed the same two other peaks that were explained by the fluctuations of the DOS near the Fermi energy. Above a given disorder strength, beyond the logarithmic decay and the peak at low $T_K \ll T_K^0$, these studies predict a non-negligible part of totally unscreened impurities, with $T_K = 0$. The disorder in the situation presented here is strong enough to observe such a feature, and it is believed that the peak at low T_K is an artifact and corresponds to these unscreened moments. Indeed, the position of the lowest peak is stuck to the magnetic field B_z, meaning that the magnetic moment is of order one for these realizations.

To confirm this statement, the density of states of the bath is further investigated. The screening mechanism is mediated by states of the bath coupled to the impurity, on an energy scale of order T_K. If no states is available in this energy scale around the Fermi energy of the electronic bath ($\epsilon = 0$ here), the screening can not occur properly and the magnetic moment is not quenched. Let $\gamma_\alpha = \sum_{i=0}^{N-1} P_{\alpha i} c_i$ be an eigenstate of the bath without the impurity with an energy E_α , the local density of states of the first site of the chain (the one that is tunnel coupled to the impurity in the full problem) is defined as:

$$\rho(\epsilon) = \sum_{\alpha=1}^{N} |P_{\alpha,0}|^2 \, \delta\left(\epsilon - E_\alpha\right), \tag{5.10}$$

and each delta distribution is enlarged by a Lorentzian function, whose width is equivalent to the level spacing, to obtain a smooth function. In Fig. 5.9, this quantity is averaged for specific realizations of the disorder: a first average is done for disorders associated to the *unscreened* moments with $T_K \ll T_K^0$, and another for *unaffected*

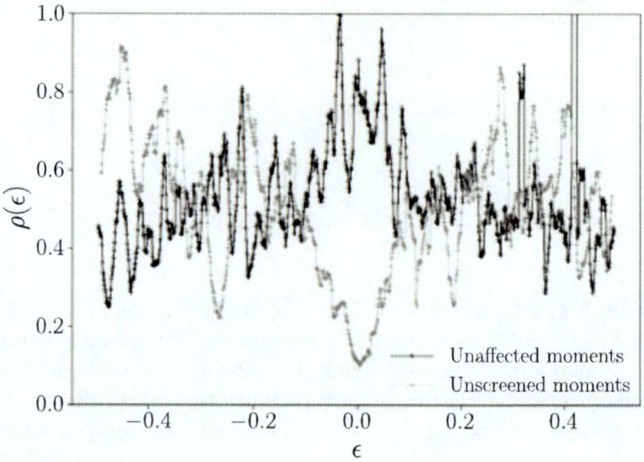

Fig. 5.9 Averaged DOS of the first site of the chain isolated from the impurity ($V = 0$). The two curves are associated to averages over different specific realizations of the disorder, related either to the full system where $T_K \ll T_K^0$ (unscreened moments) or $T_K \simeq T_K^0$ (unaffected moments)

moments with $T_K \simeq T_K^0$. The averaged DOS for unscreened moment exhibits a dip around $\epsilon \simeq 0$, while it remains large for unaffected moments, confirming the picture explained above.

In Ref. [11], the distribution of Kondo temperatures were investigated in the SIAM for different number of channels N_{ch}, and we discuss only the case $N_{ch} = 1$ corresponding to our model. On the contrary to the other study, a part of unscreened moment is not predicted in Ref. [11] , but do not observe either the three peaks structure that is found here. Their study is based on the relation between T_K and the distribution of effective width of the level Γ for each realization, and computing the moments of the distribution, they found badly behaved distribution for small number of channels, noticing that their result may be accurate for $N_{ch} \geq 10$.

To conclude, the distribution of Kondo temperatures may be more complicated than previously predicted, and the presence of a significant part of unscreened impurities is confirmed by our study, related to a disorder-induced depletion of the density of states of the bath about the Fermi level. The distribution is found to be trimodal, with one mode near the bare T_K^0, a second mode at $T_K \gg T_K^0$ reflecting the fluctuations of the charge, and an additional dominant mode at exponentially smaller T_K accounting for the depletion of the DOS near the Fermi energy. However, this study is not complete and the mechanism behind each situations must be further investigated, by varying the disorder, the interaction strength U etc. As the RGNO simulates systems with a finite size N, it should be carefully checked that these observations are not due to finite size effects and stay valid in the thermodynamic limit.

References

1. Anderson PW (1958) Absence of diffusion in certain random lattices. Phys Rev 109:1492–1505
2. Mello P, Kumar N (2004) Quantum transport in mesoscopic systems. University Press, Oxford. ISBN: 0198525826
3. Imry Y, Tinkham M (1997) Introduction to mesoscopic physics. University Press, Oxford
4. Beenakker CWJ (July1997) Random-matrix theory of quantum transport. Rev Modern Phys 69:731–808
5. Müuller CA, Delande D (2010) Disorder and interference: localization phenomena. arXiv:1005.0915
6. Dorokhov ON (1982) Transmission coefficient and the localization length of an electron in N bound disordered chains. Soviet J Exper Theor Phys Lett 36:318
7. Cornaglia PS, Grempel DR, Balseiro CA (2006) Universal distribution of kondo temperatures in dirty metals. Phys Rev Lett 96:117209
8. Dobrosavljević V, Kirkpatrick TR, Kotliar BG (1992) Kondo effect in disordered systems. Phys Rev Lett 69:1113–1116
9. Miranda E, Dobrosavljevic V, Kotliar G (1997) Disorder-driven non-Fermi liquid behavior in Kondo alloys. Phys Rev Lett 78:290–293. ISSN: 0031-9007, 1079-7114
10. Kettemann S, Mucciolo ER (2006) Free magnetic moments in disordered metals. J Exper Theor Phys Lett 83:240–245. ISSN: 0021- 3640, 1090-6487
11. Kaul RK, Ullmo D, Baranger HU (2003) Mesoscopic fluctuations in quantum dots in the Kondo regime. Phys Rev B 68:161305. ISSN: 0163-1829, 1095-3795

Chapter 6
Random Matrix Impurity Model

6.1 Motivations

The previous chapter was devoted to the full many-body description of correlations in a disordered host surrounding an impurity. The correlation properties, that can be quantified through the given Kondo temperature, revealed to be mostly related to local quantities, especially the occupation of the impurity level. This can be understood because the non-triviality of the problem comes from the dynamic of the d-level, which becomes frozen if the impurity is fully occupied or empty. To understand the bimodality of the distribution of T_K induced by the plateau in $P(n_d)$ around $n_d = 1/2$ in the IRLM, the free problem $U = 0$ is further investigated in this chapter. The importance of local properties suggests that the microscopic details of the disordered environment may not be crucial, and that underlying mechanisms could be qualitatively explained by a simpler model, into which the bath part is defined randomly. Numerical and analytical results are both presented in this chapter, and connections with the microscopic model presented in the previous chapter appear through the distribution function $P(n_d)$.

The benefit of considering a totally random one-body part for the Hamiltonian is the access to the machinery of the random matrix theory, which contains known analytical formulations. The theory concerning random matrices was introduced in the context of nuclear physics in the last century to analyse the structure of excited states of heavy nuclei, that was too complex to describe microscopically [1–3]. However, this complexity was on the contrary a boon: they are statistical in nature, and only averaged properties need to be explained. On the contrary to statistical mechanics, which assumes that all states of a large ensemble are equally probable, random matrix theory describes systems whose nature (its hamiltonian) is unknown, not its states. This means in other words that every mechanism of interaction, scattering, potential, etc. is equally probable [4]. The accuracy of this description led to the development of theories concerning eigenvalues and eigenvectors of matrices with random entries [4–7], that found applications in a context much wider than nuclear

M. Debertolis, *Quantum Impurity Problems in the Framework of Natural Orbitals*, Springer Theses, https://doi.org/10.1007/978-3-031-47233-6_6

resonances, from biology [8] to finance [9]. For general reviews concerning random matrix theory, and its applications to condensed matter physics, the reader can refer to Refs. [10–12].

However, models of disordered quantum impurities were not as extensively studied with random matrices as the rest. Studies have been mostly focused on the distribution of Kondo temperatures [13–15], which is the relevant scale at low temperatures, and can be related to the distribution of certain physical quantities of interest. These studies were based on scaling arguments for the Kondo temperature, and the wave functions, for instance, are out of the scope of such methods. A later work concerning the *mesoscopic* Anderson impurity problem [16] used RMT beyond the distribution of T_K, and calculated the joint distribution of eigenvalues of the problem with and without the magnetic impurity. A simple toy model based on these result was proposed, that provides simple expression for spectral correlations in the impurity model. Nevertheless, only spectral properties were analyzed, and properties related to the eigenstates remained unknown for the full problem at any coupling between the lead and the impurity. In another context, a model similar to the one that we will discuss hereafter was studied to describe Mie scattering of light by large large grains [17, 18].

The study that is proposed in this chapter aims to unify the RMT description of spectrum and states, with the calculation of the joint probability of energies and eigenstates amplitude of the full impurity problem, in the non-interacting limit. The non-interacting limit can be extended as usual at a mean field level, renormalizing the parameters of the free Hamiltonian. The model is first of all introduced in its different regimes with several numerical results, which are related to the results of Chap. 5. Then, we analyse a simplified two-sites model which reproduce at least qualitatively, and in some limits quantitatively, the occupancy distribution of the impurity $P(n_d)$, with which analytical results can be easily calculated. This limit is nevertheless too simplified to capture every property, what leads us to the development of the joint probability density function (PDF) of energies and eigenstate amplitudes of the full problem, which contains information to compute any averaged quantity. We will end this chapter with large-N calculations of some quantities, starting from the PDF previously built.

6.2 Random Matrix Resonant Level Model

As was introduced above, the model that we are going to study here needs to capture the local effects on an impurity coupled to a disordered media, that is modeled by a bath whose properties are totally unknown: the one-body part of the Hamiltonian concerning the bath is a matrix with gaussian random entries, provided that it is hermitian. This Hamiltonian reads:

$$\hat{H}_{\text{imp}} = \epsilon_d d^\dagger d + V\left(d^\dagger c_1 + \text{H.c.}\right) + \sum_{i,j=1}^{N} G_{ij} c_i^\dagger c_j, \tag{6.1}$$

where ϵ_d is the chemical potential of the impurity level, V the tunneling rate to a single site of the bath, and $\hat{G}$ a hermitian matrix with as entries independent random variables following a centered Gaussian distribution:

$$P\left(G_{ij}\right) = \frac{1}{\sqrt{2\pi}\sigma} \exp\left(-\frac{1}{2\sigma^2} G_{ij}^2\right), \tag{6.2}$$

with variance σ. This model has hence two energy scales, σ that mimics the strength of the disorder, and V for the tunneling rate between the impurity and the bath. Note that the impurity level is totally deterministic, and this will be important in the limit we are interested in: if disorder is added on the impurity, the limit $V/\sigma \to 0$ does not exist as V would be dominated by σ on the impurity. The model would become equivalent to the free problem, and no significant effect would be expected in this limit.

Concerning the matrix $\hat{G}$, that corresponds to the *free* problem (to be compared to the *perturbed* problem, represented by $\hat{H}_{\text{imp}}$), it is known that its eigenvalues e are, for N large enough, distributed according to the so-called Wigner's semi-circle law [10]:

$$W(e) = \begin{cases} \frac{1}{2\pi\sigma^2}\sqrt{2N\sigma^2 - e^2}, & |e| < \sqrt{2N}\sigma, \\ 0, & |e| > \sqrt{2N}\sigma. \end{cases} \tag{6.3}$$

With this definition, the integral of W on the real axis is equal to the number of states N, and one may normalize W to work with PDF. For the perturbed Hamiltonian, the PDF of the density of energies E was calculated in Ref. [17]. They consider a model with m Mie scatterers, and the particular case $m = 1$ corresponds to Eq. (6.1). They use the heavy formalism for large-N calculations in RMT [19–21] (Hubbard-Stratonovich transformation, supersymmetric formalism, steepest descent) that allows to find spectral properties. The averaged density of level reads:

$$\overline{\rho(E)} = \frac{1}{N+1}\left(W(E) + \frac{1}{N}\frac{W(0)\Gamma/\pi}{E^2 + (W(0)\Gamma)^2}\right), \tag{6.4}$$

which is the distribution of the free problem with an additional level, that is distributed according Lorentzian of width $W(0)\Gamma$, with $\Gamma = V^2/\sigma^2$. It is clear that in this situation, the effect of the impurity is of order $1/N$ on the bath, which is explicitely revealed by this distribution function.

It is recalled that the quantity that needs to be calculated to compare to the microscopic impurity model is $P(n_d)$, a quantity that involves both the energies and the wave function, which is not given by previous studies. Hence, exact numerical results are first investigated, to clearly understand the different situations that occur when we vary the ratio V/σ. The regimes that will be analyzed are the following: $V/\sigma < \delta e$, where δe is the spacing between the first negative and the first positive level of the free problem; $V/\sigma \propto \sqrt{N}$, where the impurity is well diluted in the bath provided that $V^2 < 2N\sigma^2$, and $V >\propto \sqrt{2N}\sigma$, when bound states begin to be exist out of the band. In each of these situations, Fig. 6.1 reports the corresponding averaged distribution of energy levels and of the occupation of the localized state, defined for each realization of the random matrix as:

$$n_d = \langle d^\dagger d \rangle = \sum_{E_\alpha < 0} z_\alpha = \sum_{E_\alpha < 0} |\psi_d(\alpha)|^2, \tag{6.5}$$

where z_α is the squared amplitude of the eigenstate of energy E_α on the impurity level. Unless otherwise stated, the parameters taken for the numerical simulations in the following are $\sigma = \sqrt{2}$ and $N = 300$.

6.2.1 *Weak Coupling Regime:* $V/\sigma \sim 1/\sqrt{2N}$

This regime corresponds to the weak coupling limit $V/\sigma \to 0$. When $V/\sigma = 0^+$, the impurity is very slightly coupled to its environment, and their mutual effect is small: the amplitude of the wave function on the impurity is close to one ($z_\alpha \simeq 1$), and the density of eigenvalues of the bath is the free one. The energy level fluctuates slightly around $\epsilon_d = 0$ in a range V^2. In this extreme limit, the impurity is polarized, and n_d is always close to 0 or 1, depending on the sign of the energy level that is the closest to the Fermi level $E_0 = \min_\alpha(|E_\alpha|)$, since chosing $\epsilon_d = 0$ allows sign changes for E_0.

In the left top and bottom panels of Fig. 6.1, the averaged energy density and the distribution $P(n_d)$ are plotted, for $V = 0.5\sigma/\sqrt{2N}$. The bath states remains the ones of the free problem, given by the semi-circle, and the additional state lies at $E \simeq 0$, whose structure is well described by the Lorentzian of Eq. (6.4) in Fig. 6.2.

The corresponding $P(n_d)$ has two distinct effects: The two peaks around 0 and 1 correspond to an isolated impurity with respectively a negative or positive energy E_0, and the plateau linking the two peaks, corresponding to realizations into which the impurity starts to be diluted in the bath, and some z_α are of the same order that z_0. If V/σ is decreased, the plateau becomes smaller and smaller up to the limit $V = 0^+$.

This regime describes the same $P(n_d)$ as the one that is seen in the microscopic study of Chap. 5: the plateau in $P(n_d)$ corresponds to the crossover between the weak coupling $V/\sigma \ll 1$ and the strong coupling limit $V/\sigma \gg 1$, and as the disorder increases at fixed V, the impurity can only be polarized as it is can not couple to any states of the bath.

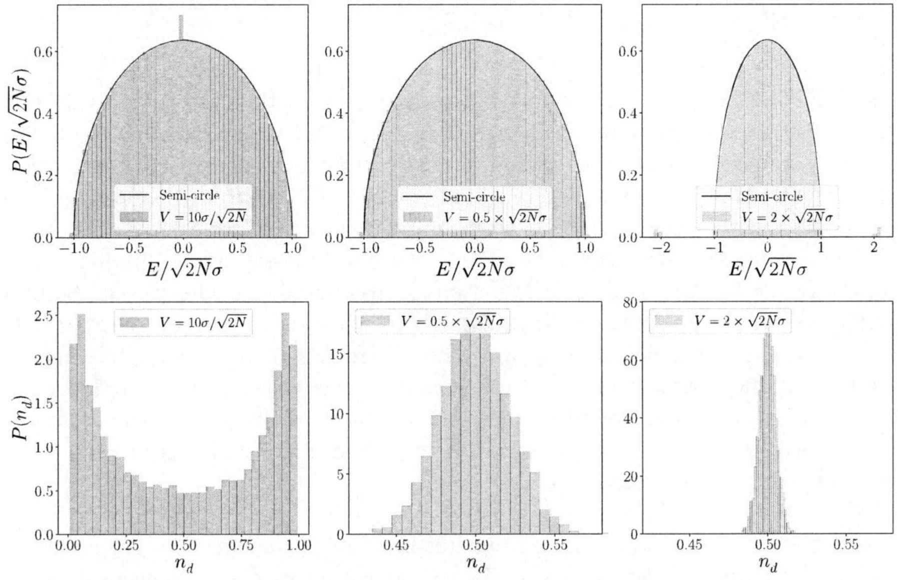

Fig. 6.1 **Top panels**: Density of energy levels for three different values of the ratio V/σ corresponding to three different regimes of the problem, for $N = 300$ (299^2 random variables). From left to right: weak coupling regime, diluted regime, bound states regime. The semi-circle law corresponding to the free problem is plotted for comparison. **Lower panels**: Distribution of the occupation of the impurity level n_d in the same three situations

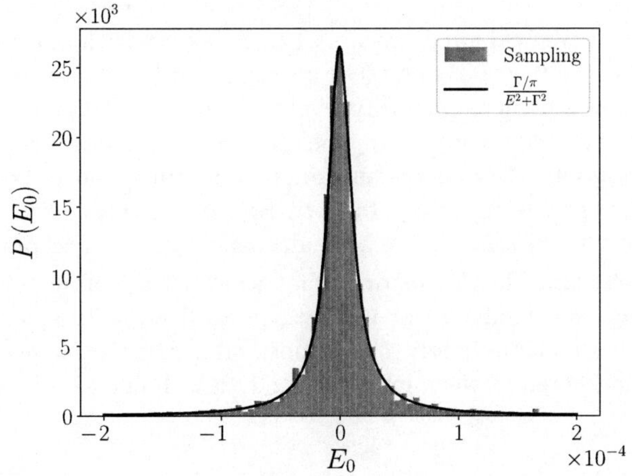

Fig. 6.2 Distribution of the closest energy level to the Fermi energy for $N = 300$ and $V/\sigma = 0.5/\sqrt{N}$. This level has the largest weight on the impurity for this value of V/σ, and its distribution is compared to the expected Lorentzian function of width $\Gamma = V^2$

6.2.2 Diluted Regime: $V/\sigma < \sqrt{2N}$

As V/σ increases from the previous situation, there a smooth crossover to what is called here the diluted limit, which is the begining of the strong coupling limit. The impurity couples to more and more states, and the amplitude of eigenstates are diluted into more and more z_α. When every z_α is of order $1/N$, the strong coupling limit is reached. However, the perturbation by the impurity is also of order $1/N$, such that it acts only as an additional state diluted in the bath and the effect is negligible.

The results of the numerical simulations are plotted the middle top and bottom panels of Fig. 6.1. The spectrum is equivalent to the free one, the effect of order $1/N$ being already negligible for $N = 300$. The distribution $P(n_d)$ is simply centered around $1/2$: there are as many levels positive and negative (up to one extra level due to the impurity), and every state has the same weight on the impurity. The fluctuations comes from occasional asymmetries of the weight of the wave function on the left and right parts of the spectrum.

This strong coupling regime corresponds in the microscopic model with weak disorder at particle-hole symmetry, where the impurity fluctuations are maximal and the impurity occupation is slightly deviated from its clean expectation value $n_d = 1/2$.

6.2.3 Bound States Regime: $V/\sigma > \sqrt{2N}$

The last situation that is explored appears when V/σ exceeds the bandwidth $\sqrt{2N}$. If $V/\sigma \to \infty$, it is clear that two bound state are separated from the system, with energy $\pm V$, with equal weight on the impurity level $z_+ = z_- = 1/2$. For $V/\sigma > \sqrt{2N}$, the picture is similar: two states are leaving the semi-circle, but the influence of the band on these states deviates their energy and amplitude from V and $1/2$ respectively.

In the top right panels of Fig. 6.1, these two *outliers* are cleary seen outside of the main band, with energies $|E_\pm| \simeq V$, while the band itself stays accurately described by the semi-circle law. The distribution of n_d is sharper as before, what is expected since fluctuations are smaller when V increases for fixed σ. The mechanism fixing n_d around $1/2$ is simpler as before, and is mostly due to the negative outlier that has almost half of the weight of the impurity $z_- \simeq 1/2$, with fluctuations induced by the band.

6.2.4 Participation Ratio

The three regimes presented above can be characterized quantitatively by the participation ratio (PR), defined as:

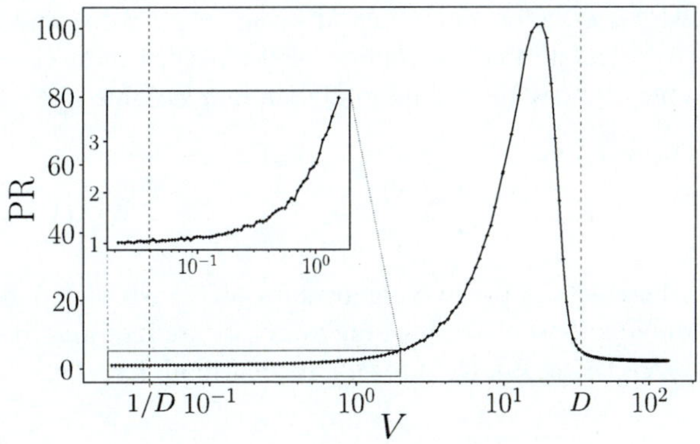

Fig. 6.3 Participation ratio as a function of V, from weak to strong coupling. The energy scales D and $1/D$ shown to delimitate the different regimes. The higher is the PR, the more diluted is the impurity in the bath

$$\mathrm{PR} = \left(\sum_{\alpha=1}^{N} z_\alpha^2\right)^{-1} = \left(\sum_{\alpha=1}^{N} |\psi_d(\alpha)|^4\right)^{-1}, \tag{6.6}$$

which counts the number of dominant eigenstates, for each realization of the random matrix. This quantity is reported with respect to V in Fig. 6.3, each point being averaged over 200 realizations of the random matrix. When V is small, the impurity is almost isolated and the PR equals 1 ($z_0 \simeq 1$). When V increases above the energy of the first level of the bath (its energy is of order $1/D$), more and more states are populated, and the PR keeps increasing until it reaches the edge of the band. After passing the edge, the PR drastically falls down to 2, indicating the presence of the two outliers, whose weight is about $1/2$ each.

Although mechanisms behind each regime seem to be driven by complicated hybridization of the impurity in the bath, it turns out that the distribution of the occupation $P(n_d)$ can be sketched by simpler arguments. Until now, the numerical results were shown for fixed $N = 300$, in order to be in the regime of large N. Actually, our numerical simulations show that $P(n_d)$ is very little changed when N is decreased, and a surprising result was that it is still qualitatively well reproduced for $N = 2$. This simplified model, for which simple analytical results can be built, is investigated now, as well as its limitations to reproduce the full system.

6.3 Toy Model for $P(n_d)$

The toy model is defined by the Hamiltonian (6.1) with $N = 2$:

$$H_2 = V\,(d^\dagger c_1 + \mathrm{H.c.}) + \epsilon_1 c_1^\dagger c_1 = \begin{pmatrix} 0 & V \\ V & \epsilon_1 \end{pmatrix}, \tag{6.7}$$

with ϵ_1 a random potential following a normal distribution $P(\epsilon_1)=(\sqrt{2\pi}\sigma)^{-1}e^{-\epsilon_1^2/2\sigma^2}$. To calculate the distribution of the occupation of the d-level, the eigenvalues and eigenvectors are related to the random variable ϵ_1:

$$\lambda_\pm = \frac{\epsilon_1}{2} \pm \frac{1}{2}\sqrt{\epsilon_1^2 + 4V^2}, \quad \psi_\pm = \frac{1}{(V^2+\lambda_\pm^2)^{1/2}}\begin{pmatrix} V \\ \lambda_\pm \end{pmatrix}, \tag{6.8}$$

where $\lambda_\pm$ are the two eigenvalues of H_2 associated to eigenvectors $\psi_\pm$. To find n_d, only the first element of eigenvectors are required, which is related to the impurity level. Using Eq. (6.5) for n_d, its distribution reads:

$$\begin{aligned} P(n_d) &= \int_{-\infty}^{+\infty} \mathrm{d}\epsilon_1 P(\epsilon_1)\,\delta\left(n_d - \sum_{n=+,-}\Theta(-\lambda_n(\epsilon_1))z_{0n}(\epsilon_1)\right) \\ &= \int_{-\infty}^{0} \mathrm{d}\epsilon_1 P(\epsilon_1)\,\delta\left(n_d - z_{0+}(\epsilon_1)\right) + \int_{0}^{+\infty} \mathrm{d}\epsilon_1 P(\epsilon_1)\,\delta\left(n_d - z_{0-}(\epsilon_1)\right), \end{aligned} \tag{6.9}$$

the second line used $\Theta(-\lambda_-(\epsilon_1)) = \Theta(\epsilon_1)$, $\Theta(-\lambda_+(\epsilon_1)) = \Theta(-\epsilon_1)$, that comes directly from Eq. (6.8). To compute the two integrals, the usual property of the delta distribution is used:

$$\int_{-\infty}^{+\infty} \mathrm{d}x\ f(x)\delta(g(x)) = \sum_{x_i}\frac{f(x_i)}{|g'(x_i)|}, \tag{6.10}$$

where x_i are the zeros of $g(x)$. In the present case, $g(x)$ has two zeros with opposite sign, and only one of them appears in the integrals. The derivative of g and its root read:

$$|g'(x)| = \frac{2V^2}{\left(4V^2+x^2\right)^{\frac{3}{2}}}, \quad x_1 = V\frac{1-2n_d}{(n_d(1-n_d))^{\frac{1}{2}}}, \tag{6.11}$$

and combining the two integrals in Eq. (6.9) leads to the desired distribution of n_d:

$$P(n_d) = \frac{V}{2\sigma\sqrt{2\pi}}\exp\left(-\frac{2V^2}{\sigma^2}\frac{(n_d-\frac{1}{2})^2}{n_d(1-n_d)}\right)\left(n_d(1-n_d)\right)^{-\frac{3}{2}}. \tag{6.12}$$

This equation shows the competition of two terms in the different regimes $V/\sigma \ll 1$ and $V/\sigma \gg 1$. If the coupling is weak, $P(n_d) \sim (n_d(1-n_d))^{-\frac{3}{2}}$, and the symmetry under exchange of n_d by $1-n_d$ gives power laws around 0 and 1, with long tails yielding the plateau of $P(n_d)$ around $n_d = 1/2$. The term in the exponential in $1/n_d(1-n_d)$ prevents the equation to diverge when n_d goes to 0 and 1. In the

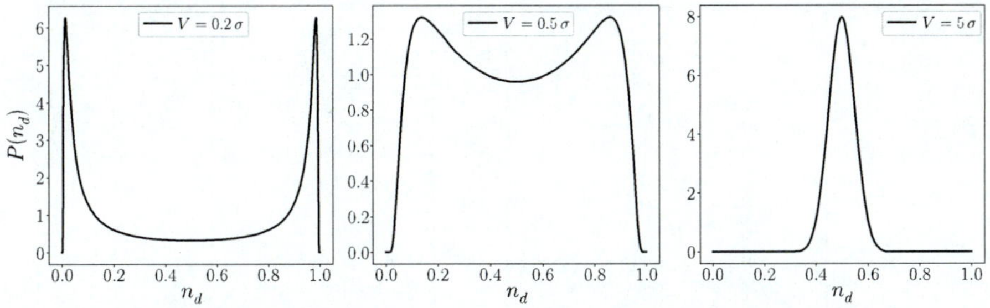

Fig. 6.4 $P(n_d)$ calculated with the analytical results of the toy model for three representative values of the coupling V/σ. The distributions in the left and middle panels ressemble to the one of the microscopic model associated to a bimodal distribution of Kondo temperatures (see Chap. 5)

strong coupling regime, the other term is dominant and $P(n_d) \sim \exp[-\frac{2V^2}{\sigma^2}(n_d - \frac{1}{2})^2]$ which is a Gaussian of mean $1/2$ and of deviation $\tilde{\sigma} = \sigma/2V$. $P(n_d)$ is plotted in these two limits in the left and right panels of Fig. 6.4 respectively. In the middle panel, an intermediate value of V/σ is plotted to show the crossover between the two regimes.

This model has however no information concerning the bath, such that the bandwidth $\sqrt{2N}\sigma$ can not be felt. In the model with N large, $P(n_d)$ is not sensible to the bandwidth, only the scale V/σ matters, which explains the success of the toy model to describe this quantity in all regimes. Nevertheless, this model is not totally quantitative, and any quantity related to the details of the wave function can not be recovered. In Fig. 6.5, the plots of Fig. 6.4 are compared to numerical simulation with $N = 300$ to investigate the deviations of the toy model. For $V/\sigma \ll 1$, the power law in $(n_d(1 - n_d))^{-3/2}$ is correct, but the edges of the distribution for $n_d \simeq 0, 1$ are not quantitatively described. This is reflected in the middle panel for the intermediate regime, the edges are enhanced in the problem with $N = 300$, so that a bigger dip

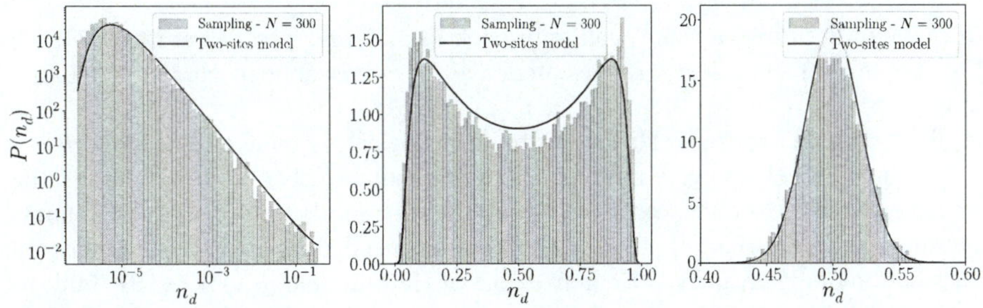

Fig. 6.5 Difference between the sampling of the random impurity model at finite $N = 300$ and the analytical results calculated for the toy model. Quantitative deviations are seen at the edge for the weak and intermediate regimes. The left panel is plotted in log-log scale to reveal the power law behavior of $P(n_d)$ at $n_d \simeq 0$, which is captured by the toy model. For V/σ large, $P(n_d)$ is accurately described as the two sites model becomes exact

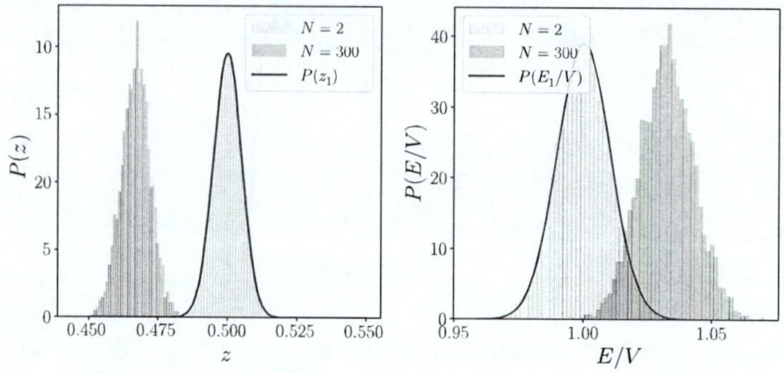

Fig. 6.6 Clear limitations of the toy model for other quantities than $P(n_d)$, here $P(z_0)$ and $P(E_0)$, the respective amplitude and energy of the positive outlier in the bound states regime. Here, $V/\sigma = 2\sqrt{2N}$, and the effects of the coupling between the impurity and the band are totally missed by the toy model

is induced around $n_d \simeq 1/2$. On the opposite, there is a quantitative agreement for $V/\sigma \gg 1$.

It is clear that in the weak coupling regime, any quantity other than $P(n_d)$ can not be obtained by this toy model. It is also the case in the strong coupling limit: when $N = 300$, for $V > \sqrt{2N}\sigma$, the two outliers could play the role of the two-sites system. But the deviations coming from the interaction with the bath renormalize the energy and the amplitude of these states. In the toy model, these distributions read:

$$P(\lambda_\pm) = \frac{1}{\sigma\sqrt{2\pi}} \exp\left(-\frac{1}{2\sigma^2}\frac{(\lambda_\pm^2 - V^2)^2}{\lambda_\pm^2}\right)\left(1 + \left(\frac{\lambda_\pm}{V}\right)^2\right). \tag{6.13}$$

$$P(z_\pm) = \frac{V}{2\sigma\sqrt{2\pi}} \exp\left(-\frac{2V^2}{\sigma^2}\frac{(z_\pm - \frac{1}{2})^2}{z_\pm(1 - z_\pm)}\right) z_\pm^{-\frac{5}{2}} \left(1 - z_\pm\right)^{-\frac{1}{2}}, \tag{6.14}$$

from which it turns out that λ can only stick to $\pm V$, and z to $1/2$ at large V/σ. In Fig. 6.6, this effect is clearly represented, and the curves are expected to match only if $V \to \infty$.

To sum up, this toy model brought simple analytical formulas that describe $P(n_d)$ at least qualitatively in every regime, and even quantitatively in some situations, that was unexpected. However, other local quantities are poorly reproduced or can not be captured, so that a solution for the full problem is needed. In the following, the joint distribution of eigenvalues and eigenvectors of Hamiltonian (6.1) is exactly built for any value of N.

6.4 Full PDF of the Perturbed Problem

The work presented in this part aims to find the joint probability distribution of the set of eigenvalues and eigenvectors of the random impurity model. The starting point is the distribution of these variables in the free model, that follows the so-called Porter-Thomas distribution [2]:

$$P_{\text{P-T}}(\{e_\alpha\},\{r_\alpha\}) = \prod_{1\leq\alpha<\beta}^{N} |e_\alpha - e_\beta| \frac{1}{\sqrt{r_\alpha}} \delta\left(\sum_{\beta=1}^{N} r_\beta - 1\right) e^{-\frac{1}{2\sigma^2}\sum\limits_{\beta=0}^{N} e_\beta^2}, \tag{6.15}$$

where $r_\alpha = |\phi_i(\alpha)|^2$ is the square of any element i of eigenvector α that holds an energy e_α. The δ distribution imposes the wave function to be normalized. A similar study was performed for a random model perturbed by a local perturbation [22, 23]. In their model, the perturbation is a deterministic potential on a given entry of the random matrix, corresponding to a rank-one perturbation of the random matrix of rank N. In our system, the perturbation is of rank two, and the matrix problem (6.1) can be reformulated in a more convenient way for this study:

$$H_{ij} = G_{ij} + V(\delta_{i0}\delta_{j1} + \delta_{i1}\delta_{j0}), \quad \text{with} \quad \begin{cases} G_{ij} = 0, & \forall i = 0 \text{ or } j = 0 \,|\, i \neq j \\ G_{ij} = \epsilon_d, & \text{for } \; i = j = 0, \\ G_{ij} = G_{ij} & \text{otherwise.} \end{cases} \tag{6.16}$$

The entries G_{ij} for $i, j > 0$ are distributed following a Gaussian distribution of zero mean and of deviation σ, and are real and symmetric. The ensemble under consideration is known as the Gaussian orthogonal ensemble (GOE), named after the orthogonal matrix that diagonalizes hermitian matrices. The joint distribution of the eigenvalues and eigenvectors of $\hat{G}$ is known and corresponds to Eq. (6.15), and the constraints required in Eq. (6.16) can be introduced as:

$$P_G(\{e_\alpha\},\{r_\alpha\}) = P_{\text{P-T}}(\{e_\alpha\},\{r_\alpha\})\delta\left(e_0 - \epsilon_d\right)\delta\left(r_0\right). \tag{6.17}$$

We are looking the distribution of the occupation of the impurity $P(n_d)$, defined as $n_d = \sum_{E_\alpha<0} |\psi_0(\alpha)|^2$, where $\psi_0(\alpha)$ is the site 0 component of the eigenvector α of matrix $\hat{H}$, and E_α is its associated eigenvalue. In the random impurity model H, if we call $z_\alpha = |\psi_0(\alpha)|^2$ the amplitude of the eigenvector of energy E_α such that the desired function is $P(\{E_\alpha\},\{z_\alpha\})$. Hence, one wants to perform the change of variables between the $(N-1)^2$ variables of the free problem to the $(N-1)^2$ ones of the perturbed problem. There are however N^2 entries in H, but since the first line and the first column are known, they are not random variables and are fixed by two constraints imposed on the eigenvalues and eigenvectors of H. The details of this calculation are presented below, but the reader can go to Eq. (6.76) to skip the explicit derivation.

6.4.1 Definition of the Eigenvalue Problem

For both matrices $\hat{H}$ and $\hat{G}$, their respective eigenvalues and eigenvectors can be written with the following equations:

$$\begin{cases} \sum_{j=0}^{N} G_{ij}\phi_j(\alpha) = e_\alpha \phi_i(\alpha), \quad \text{with} \quad e_0 = \epsilon_d, \; \phi_0(\alpha) = \delta_{0\alpha} \\ \sum_{j=0}^{N} H_{ij}\psi_j(\alpha) = E_\alpha \psi_i(\alpha). \end{cases} \tag{6.18}$$

The orthonormality of eigenvectors ψ and ϕ is defined as usual:

$$\begin{cases} \sum_{i}^{N} \phi_i(\alpha)\,\phi_i(\beta) = \delta_{\alpha\beta}, \\ \sum_{\alpha}^{N} \phi_i(\alpha)\,\phi_j(\alpha) = \delta_{ij}. \end{cases} \tag{6.19}$$

The notation $\phi_i(\alpha)$ (and equivalently for $\psi_i(\alpha)$) is used in the following, and corresponds to the ith component of the αth eigenvector which holds an energy e_α (E_α). The transformation matrix $\hat{C}$, which changes the *new* basis ψ to the *old* one ϕ, is defined as:

$$\begin{cases} \psi_j(\alpha) = \sum_{\beta}^{N} C_{\alpha\beta}\,\phi_j(\beta) \\ \phi_j(\alpha) = \sum_{\beta}^{N} C^{-1}_{\alpha\beta}\,\psi_j(\beta) \end{cases}, \quad \text{with} \quad C^{-1}_{\alpha\beta} = C_{\beta\alpha} \quad (\hat{C} \text{ orthogonal}). \tag{6.20}$$

The problem is now completely defined, and explicit relations must be found between the old set of variables and the new ones to express the problem in the eigenbasis of the Hamiltonian $\hat{H}$. The PDF being defined through integrals of the old variables, the determinant of the Jacobian defined for the change of basis is also calculated in the following. The change of basis is now presented step by step in the spirit of Ref. [23], which developed the calculation for a perturbation of rank 1.

6.4.2 Change of Basis

6.4.2.1 From Eigenvectors $\{\phi_1(\alpha)\}$ to Eigenvalues $\{E_\alpha\}$

The desired change of variables $\left(\{e_\alpha\}, \{r_\alpha\}\right) \to \left(\{E_\alpha\}, \{z_\alpha\}\right)$ can not be done in one shot, and hence will be done in two steps. In this first section, eigenvectors of the free problem will be expressed through the eigenvalues of the perturbed problem, this choice being relevant given the form of the equations. Firstly, elements of the transformation matrix $\hat{C}$ are expressed in an explicit form:

$$\begin{aligned}
&\sum_{j=0}^{N} H_{ij}\psi_j(\alpha) = E_\alpha \psi_i(\alpha) \quad \Rightarrow \quad \sum_{j=0}^{N}\Big[G_{ij} + V(\delta_{i0}\delta_{j1} + \delta_{i1}\delta_{j0})\Big]\psi_j(\alpha) = E_\alpha \psi_i(\alpha) \\
&\Rightarrow \sum_{\beta=0}^{N} C_{\alpha\beta}\left(\left[\sum_{j=0}^{N} G_{ij}\phi_j(\beta)\right] - E_\alpha \phi_i(\beta)\right) = -V\sum_{j=0}^{N}(\delta_{i0}\delta_{j1} + \delta_{i1}\delta_{j0})\psi_j(\alpha) \\
&\Rightarrow \sum_{\beta=0}^{N} C_{\alpha\beta}\left(E_\alpha - e_\beta\right)\phi_i(\beta) = V\left(\delta_{i1}\psi_0(\alpha) + \delta_{i0}\psi_1(\alpha)\right) \\
&\Rightarrow \sum_{i,\beta=0}^{N} C_{\alpha\beta}\left(E_\alpha - e_\beta\right)\phi_i(\gamma)\,\phi_i(\beta) = V\sum_{i=0}^{N}\phi_i(\gamma)\left(\delta_{i1}\psi_0(\alpha) + \delta_{i0}\psi_1(\alpha)\right) \\
&\Rightarrow C_{\alpha\gamma}\left(E_\alpha - e_\gamma\right) = V\left(\phi_1(\gamma)\,\psi_0(\alpha) + \phi_0(\gamma)\,\psi_1(\alpha)\right) \\
&\qquad\qquad \Rightarrow \boxed{C_{\alpha\beta} = V\frac{\phi_1(\beta)\,\psi_0(\alpha) + \phi_0(\beta)\,\psi_1(\alpha)}{E_\alpha - e_\beta}}.
\end{aligned} \tag{6.21}$$

This last equality is injected in the change of basis (6.20):

$$\begin{cases}
\sum\limits_{\beta=0}^{N} C_{\alpha\beta}\phi_0(\beta) = \sum\limits_{\beta=0}^{N} \frac{V}{E_\alpha - e_\beta}\left[\phi_1(\beta)\,\phi_0(\beta)\,\psi_0(\alpha) + |\phi_0(\beta)|^2\psi_1(\alpha)\right], \\
\sum\limits_{\beta=0}^{N} C_{\alpha\beta}\phi_1(\beta) = \sum\limits_{\beta=0}^{N} \frac{V}{E_\alpha - e_\beta}\left[\phi_0(\beta)\,\phi_1(\beta)\,\psi_1(\alpha) + |\phi_1(\beta)|^2\psi_0(\alpha)\right], \\
\phi_0(\alpha) = \delta_{\alpha 0}, \quad \text{by definition,}
\end{cases}$$

$$\Rightarrow \begin{cases}
\psi_0(\alpha) = \sum\limits_{\beta=0}^{N} \frac{V}{E_\alpha - e_\beta}\left(\delta_{\beta 0}\phi_1(\beta)\,\psi_0(\alpha) + \delta_{\beta 0}\psi_1(\alpha)\right), \\
\psi_1(\alpha) = \sum\limits_{\beta=0}^{N} \frac{V}{E_\alpha - e_\beta}\left(\delta_{\beta 0}\phi_1(\beta)\,\psi_1(\alpha) + |\phi_1(\beta)|^2\psi_0(\alpha)\right),
\end{cases} \tag{6.22}$$

$$\Rightarrow \begin{cases}
\psi_0(\alpha)\left(1 - \frac{V\phi_1(0)}{E_\alpha - \epsilon_d}\right) = \frac{V}{E_\alpha - \epsilon_d}\,\psi_1(\alpha), \\
\psi_1(\alpha)\left(1 - \frac{V\phi_1(0)}{E_\alpha - \epsilon_d}\right) = \sum\limits_{\beta=0}^{N} \frac{V}{E_\alpha - e_\beta}|\phi_1(\beta)|^2\psi_0(\alpha),
\end{cases}$$

$$\Rightarrow \begin{cases}
\psi_0(\alpha)\left(1 - \frac{V\phi_1(0)}{E_\alpha - \epsilon_d}\right) = \frac{V}{E_\alpha - \epsilon_d}\,\psi_1(\alpha), \\
\left(E_\alpha - \epsilon_d\right)\left(1 - \frac{V\phi_1(0)}{E_\alpha - \epsilon_d}\right)^2 = \sum\limits_{\beta=0}^{N} \frac{V^2}{E_\alpha - e_\beta}|\phi_1(\beta)|^2.
\end{cases}$$

At this point, $\phi_1(0) = 0$ is also used to simplify the equations above, since this variable is exactly known and will not be integrated at the end. Thus, there are $N-1$ variables on one side $(|\phi_1(1)|^2, \ldots, |\phi_1(N)|^2)$, that are expressed in N

new variables $(E_0, \ldots, E_N)$, thus one of the variables E_α is not independent. For the moment, this extra variable is not explicitely determined, and is written $E_0 = f(\{E_{\alpha\geq 1}\}, \{\psi_{i\in\{1,0\}}(\alpha)\})$. The change of variable $\phi_1\,(\alpha > 0) \to E_\alpha$ is done first, so that e_α and $\psi_{i\in\{1,0\}}(\alpha)$ are supposed to be known here. The system to solve is thus the following:

$$\begin{cases} \psi_0\,(\alpha) = \frac{V}{E_\alpha - \epsilon_d}\,\psi_1\,(\alpha)\,, \quad \forall \alpha \geq 0, \\ \boxed{(E_\alpha - \epsilon_d) = \sum_{\beta\geq 1}^{N} \frac{V^2}{E_\alpha - e_\beta} |\phi_1\,(\beta)\,|^2, \quad \forall \alpha \geq 1}. \end{cases} \tag{6.23}$$

The second line corresponds to the eigenvalues equation, for the variables $|\phi_1\,(\beta \geq 1)\,|^2$, that is to be solved. This system can be written in a matrix form:

$$\hat{A}\,\vec{b} = \vec{\Lambda}, \tag{6.24}$$

where ${}^t\vec{\Lambda} = \Big((E_1 - \epsilon_d), \ldots, (E_N - \epsilon_d)\Big)$, ${}^t\vec{b} = V^2\Big(|\phi_1\,(1)\,|^2, \ldots, |\phi_1\,(N)\,|^2\Big)$, and $\hat{A}$ a Cauchy matrix defined as $A_{ij} = \frac{1}{E_i - e_j}$. Cramer's rule is used to determine the solutions $|\phi_1\,(\beta)\,|^2$ through determinants of matrices:

$$V^2|\phi_1\,(\beta)\,|^2 = \frac{\det(\hat{A}_\beta)}{\det(\hat{A})}, \tag{6.25}$$

with $\hat{A}_\beta$ a matrix defined by the subsitution of the column β of $\hat{A}$ by $\vec{\Lambda}$:

$$A_\beta = \begin{pmatrix} \frac{1}{E_1 - e_1} & \cdots & E_1 - \epsilon_d & \cdots & \frac{1}{E_1 - e_N} \\ \vdots & \ddots & \vdots & \ddots & \vdots \\ \vdots & \ddots & \vdots & \ddots & \vdots \\ \frac{1}{E_N - e_1} & \cdots & E_N - \epsilon_d & \cdots & \frac{1}{E_N - e_N} \end{pmatrix}. \tag{6.26}$$

The determinant of $\hat{A}_\beta$ is calculated with the Laplace expansion of the βth column:

$$\det(\hat{A}_\beta) = \sum_{\alpha\geq 1}^{N} (-1)^{\alpha+\beta}(E_\alpha - \epsilon_d)\det(A_{\beta\alpha}), \tag{6.27}$$

where $A_{\beta\alpha}$ is a matrix of rank $N-2$, defined by $\hat{A}$ without its row α and its column β. The determinant of $\hat{A}$ is then also computed:

$$\det(\hat{A}) = \det\begin{pmatrix} \frac{1}{E_1-e_1} & \cdots & \frac{1}{E_1-e_N} \\ \vdots & \ddots & \vdots \\ \frac{1}{E_N-e_1} & \cdots & \frac{1}{E_N-e_N} \end{pmatrix}. \tag{6.28}$$

The column β is substracted to each column $j \neq \beta$, which leaves the determinant unchanged:

$$\frac{1}{E_i - e_j} - \frac{1}{E_i - e_\beta} = \frac{E_i - e_\beta - E_i + e_j}{(E_i - e_j)(E_i - e_\beta)} = \frac{(e_j - e_\beta)}{(E_i - e_\beta)(E_i - e_j)}. \tag{6.29}$$

Each column is factorized by the term $(e_j - e_\beta)$, and each row by $\frac{1}{E_i - e_\beta}$.

$$\det(A) = \prod_{\gamma\geq 1} \frac{1}{E_\gamma - e_\beta} \prod_{\substack{\gamma\geq 1,\\ \gamma\neq\beta}} (e_\gamma - e_\beta) \det\begin{pmatrix} \frac{1}{E_1-e_1} & \dots 1 \dots & \frac{1}{E_1-e_N} \\ \vdots & \ddots \vdots \ddots & \vdots \\ \vdots & \ddots \vdots \ddots & \vdots \\ \frac{1}{E_N-e_1} & \dots 1 \dots & \frac{1}{E_N-e_N} \end{pmatrix}. \tag{6.30}$$

Now the row α has to be eliminated: the row α is substracted to each row i except $i = \alpha$:

$$\frac{1}{E_i - e_j} - \frac{1}{E_\alpha - e_j} = \frac{E_\alpha - e_j - E_i + e_j}{(E_\alpha - e_j)(E_i - e_j)} = \frac{(E_\alpha - E_i)}{(E_\alpha - e_j)(E_i - e_j)}, \tag{6.31}$$

and as before each column except the β one is factorized by $\frac{1}{E_\alpha - e_j}$, and each row except the α one by $(E_\alpha - E_i)$:

$$\det(\hat{A}) = \frac{\prod_{\substack{\gamma\geq 1,\\ \gamma\neq\beta}} (e_\gamma - e_\beta) \prod_{\substack{\gamma\geq 1,\\ \gamma\neq\alpha}} (E_\alpha - E_\gamma)}{\prod_{\gamma\geq 1} (E_\gamma - e_\beta) \prod_{\substack{\gamma\geq 1,\\ \gamma\neq\beta}} (E_\alpha - e_\gamma)} \det\begin{pmatrix} \frac{1}{E_1-e_1} & \dots 0 \dots & \frac{1}{E_1-e_N} \\ \vdots & \ddots \vdots \ddots & \vdots \\ 1 & \ddots 1 \ddots & 1 \\ \vdots & \ddots \vdots \ddots & \vdots \\ \frac{1}{E_N-e_1} & \dots 0 \dots & \frac{1}{E_N-e_N} \end{pmatrix},$$

$$\det(\hat{A}) = \frac{\prod_{\substack{\gamma\geq 1,\\ \gamma\neq\beta}} (e_\gamma - e_\beta) \prod_{\substack{\gamma\geq 1,\\ \gamma\neq\alpha}} (E_\alpha - E_\gamma)}{\prod_{\gamma\geq 1} (E_\gamma - e_\beta) \prod_{\gamma\geq 1} (E_\alpha - e_\gamma)} (E_\alpha - e_\beta)(-1)^{\alpha+\beta} \det(\hat{A}_{\beta\alpha}). \tag{6.32}$$

Putting everything back in (6.25), the equation becomes simply:

$$|\phi_1(\beta)|^2 = \frac{1}{V^2}\frac{\prod_{\gamma\geq 1}(E_\gamma - e_\beta)}{\prod_{\substack{\gamma\geq 1,\\ \gamma\neq\beta}}(e_\gamma - e_\beta)}\sum_{\alpha\geq 1}\left[\frac{(E_\alpha-\epsilon_d)}{(E_\alpha - e_\beta)}\frac{\prod_{\gamma\geq 1}(E_\alpha - e_\gamma)}{\prod_{\substack{\gamma\geq 1,\\ \gamma\neq\alpha}}(E_\alpha - E_\gamma)}\right]. \quad (6.33)$$

However, simple arguments from complex analysis and contour integration can be used to express the following sums [24]:

$$\sum_{j=1}^{N}\frac{x_j^r}{\prod_{\substack{i=1,\\ i\neq j}}^{N}(x_j - x_i)} = \begin{cases} 0, & \text{for } 0\leq r < N-1,\\ 1, & \text{for } r = N-1,\\ \sum_{k=1}^{N} x_k, & \text{for } r = N, \end{cases} \quad (6.34)$$

such that Eq. (6.33) can be simplified:

$$\begin{aligned}\sum_{\alpha\geq 1}\left[\frac{(E_\alpha-\epsilon_d)}{(E_\alpha - e_\beta)}\frac{\prod_{\gamma\geq 1}(E_\alpha - e_\gamma)}{\prod_{\substack{\gamma\geq 1,\\ \gamma\neq\alpha}}(E_\alpha - E_\gamma)}\right] &= \sum_{\alpha\geq 1}\left[\frac{(E_\alpha-\epsilon_d)\prod_{\substack{\gamma\geq 1,\\ \gamma\neq\beta}}(E_\alpha - e_\gamma)}{\prod_{\substack{\gamma\geq 1,\\ \gamma\neq\alpha}}(E_\alpha - E_\gamma)}\right]\\ &= \sum_{\alpha\geq 1}\frac{E_\alpha^N - \left[\sum_{\gamma\geq 1}^{N}(e_\gamma - e_\beta + \epsilon_d)\right]E_\alpha^{N-1} + P(E_\alpha^{r<(N-1)})}{\prod_{\substack{\gamma\geq 1,\\ \gamma\neq\alpha}}(E_\alpha - E_\gamma)}.\end{aligned} \quad (6.35)$$

There is is a sum of 3 terms to calculate: the two first are respectively proportional to E_α^N and E_α^{N-1}, and the last one corresponds to a polynome of degree $< N-1$. Using the simplifications of Eq. (6.34), Eq. (6.33) becomes:

$$\begin{aligned}|\phi_1(\beta)|^2 &= \frac{1}{V^2}\frac{\prod_{\gamma\geq 1}(E_\gamma - e_\beta)}{\prod_{\substack{\gamma\geq 1,\\ \gamma\neq\beta}}(e_\gamma - e_\beta)}\sum_{\alpha\geq 1}(E_\alpha - e_\alpha + e_\beta - \epsilon_d)\\ \Rightarrow &\boxed{r_\beta = |\phi_1(\beta)|^2 = \frac{1}{V^2}\frac{\prod_{\gamma\geq 1}(E_\gamma - e_\beta)}{\prod_{\substack{\gamma\geq 1,\\ \gamma\neq\beta}}(e_\gamma - e_\beta)}(e_\beta - E_0)}.\end{aligned} \quad (6.36)$$

The eigenvectors of the free problem are now totally determined in terms of the old and new eigenvalues. For this change of variables, the determinant of the Jacobian is needed:

$$[\hat{J}_{\phi\to E}]_{\alpha,\beta} = \frac{\partial r_\beta}{\partial E_\alpha}. \tag{6.37}$$

The derivative for each entry of $\hat{J}_{\phi\to E}$ is calculated:

$$\begin{aligned}\frac{\partial r_\beta}{\partial E_\alpha} &= \frac{r_\beta}{E_\alpha - e_\beta} + \frac{\prod\limits_{\gamma\geq 1}(E_\gamma - e_\beta)}{V^2 \prod\limits_{\substack{\gamma\geq 1,\\ \gamma\neq\beta}}(e_\gamma - e_\beta)},\\ \Rightarrow \det\left(\hat{J}_{\phi\to E}\right) &= \det\left(\frac{r_\beta}{E_\alpha - e_\beta} + \frac{\prod\limits_{\gamma\geq 1}(E_\gamma - e_\beta)}{V^2 \prod\limits_{\substack{\gamma\geq 1,\\ \gamma\neq\beta}}(e_\gamma - e_\beta)}\right).\end{aligned} \tag{6.38}$$

The second term is identified from Eq. (6.36):

$$\frac{\prod\limits_{\gamma\geq 1}(E_\gamma - e_\beta)}{V^2 \prod\limits_{\substack{\gamma\geq 1,\\ \gamma\neq\beta}}(e_\gamma - e_\beta)} = \frac{-r_\beta}{E_0 - e_\beta}, \tag{6.39}$$

so that the determinant is straightforwardly calculated:

$$\begin{aligned}&\Rightarrow \det\left(\frac{\partial r_\beta}{\partial E_\alpha}\right) = \det\left(\frac{r_\beta}{E_\alpha - e_\beta} - \frac{r_\beta}{E_0 - e_\beta}\right) = \det\left(\frac{r_\beta(E_0 - E_\alpha)}{(E_0 - e_\beta)(E_\alpha - e_\beta)}\right)\\ &= \prod_{\gamma\geq 1}\left(\frac{E_0 - E_\gamma}{E_0 - e_\gamma}\right)\prod_{\gamma\geq 1}(r_\gamma)\det\left(\frac{1}{E_\alpha - e_\beta}\right)\\ &= \prod_{\gamma\geq 1}\left(\frac{E_0 - E_\gamma}{E_0 - e_\gamma}\right)\frac{(-1)^{N-1}}{V^{2(N-1)}}\frac{\prod\limits_{\delta,\gamma\geq 1}(E_\gamma - e_\delta)}{\prod\limits_{\substack{\delta,\gamma\geq 1,\\ \gamma\neq\delta}}(e_\gamma - e_\delta)}\prod_{\delta\geq 1}(E_0 - e_\delta)\frac{\prod\limits_{1\leq\gamma<\delta}(E_\delta - E_\gamma)(e_\gamma - e_\delta)}{\prod\limits_{\gamma,\delta\geq 1}(E_\gamma - e_\delta)}\\ &\Rightarrow \boxed{\det(\hat{J}_{\phi\to E}) = \frac{1}{V^{2(N-1)}}\prod_{1\leq\gamma<\delta}\frac{(E_\delta - E_\gamma)}{(e_\gamma - e_\delta)}\prod_{\gamma\geq 1}\left(E_\gamma - E_0\right)}.\end{aligned} \tag{6.40}$$

The free eigenvectors can thus be correctly replaced by the eigenvalues of the perturbed problem in the PDF. The free eigenvalues are now replaced by the perturbed eigenvectors, with the same procedure as above, to close the problem in the new set of variables.

6.4.2.2 From Eigenvalues $\{e_\alpha\}$ to Eigenvectors $\{\psi_0(\alpha)\}$

The orthogonality of the matrix $\hat{C}$ is used to directly compute its inverse:

$$\Rightarrow \boxed{C^{-1}_{\alpha\beta} = C_{\beta\alpha} = V\frac{\phi_1(\alpha)\,\psi_0(\beta) + \phi_0(\alpha)\,\psi_1(\beta)}{E_\beta - e_\alpha}}. \tag{6.41}$$

As before, the previous equation is reinjected in (6.20):

$$\begin{cases} \sum\limits_{\beta=0}^{N} C^{-1}_{\alpha\beta}\psi_0(\beta) = \sum\limits_{\beta=0}^{N} \frac{V}{E_\beta - e_\alpha}\left[|\psi_0(\beta)|^2\phi_1(\alpha) + \psi_0(\beta)\,\psi_1(\beta)\,\phi_0(\alpha)\right], \\ \sum\limits_{\beta=0}^{N} C^{-1}_{\alpha\beta}\psi_1(\beta) = \sum\limits_{\beta=0}^{N} \frac{V}{E_\beta - e_\alpha}\left[\psi_0(\beta)\,\psi_1(\beta)\,\phi_1(\alpha) + |\psi_1(\beta)|^2\phi_0(\alpha)\right], \\ \phi_0(\alpha) = \delta_{\alpha 0},\ \phi_1(0) = 0, \quad \text{by definition.} \end{cases} \tag{6.42}$$

The system to solve in this case is the following:

$$\begin{cases} \phi_0(\alpha) = \sum\limits_{\beta=0}^{N} \frac{V}{E_\beta - e_\alpha}\left[|\psi_0(\beta)|^2\phi_1(\alpha) + \psi_0(\beta)\,\psi_1(\beta)\,\delta_{\alpha 0}\right] \\ \phi_1(\alpha) = \sum\limits_{\beta=0}^{N} \frac{V}{E_\beta - e_\alpha}\left[\psi_0(\beta)\,\psi_1(\beta)\,\phi_1(\alpha) + |\psi_1(\beta)|^2\delta_{\alpha 0}\right] \end{cases}$$

$$\Rightarrow \begin{cases} \sum\limits_{\beta=0}^{N} \frac{V}{E_\beta - \epsilon_d}\psi_0(\beta)\,\psi_1(\beta) = 1, \quad \alpha = 0 \quad \Rightarrow \sum\limits_{\beta=0}^{N} |\psi_0(\beta)|^2 = 1, & (6.43a) \\ \sum\limits_{\beta=0}^{N} \frac{V}{E_\beta - e_\alpha}\psi_0(\beta)\,\psi_1(\beta) = 1, \quad \forall\alpha > 0 \Rightarrow \sum\limits_{\beta=0}^{N} \frac{E_\beta - \epsilon_d}{E_\beta - e_\alpha}|\psi_0(\beta)|^2 = 1, & (6.43b) \\ \sum\limits_{\beta=0}^{N} \frac{V}{E_\beta - \epsilon_d}|\psi_1(\beta)|^2 = 0, \quad \alpha = 0 \quad \Rightarrow \sum\limits_{\beta=0}^{N} \frac{E_\beta - \epsilon_d}{V}|\psi_0(\beta)|^2 = 0, & (6.43c) \\ \sum\limits_{\beta=0}^{N} \frac{V}{E_\beta - e_\alpha}|\psi_0(\beta)|^2 = 0, \quad \forall\alpha > 0 \quad \Rightarrow \sum\limits_{\beta=0}^{N} \frac{V}{E_\beta - e_\alpha}|\psi_0(\beta)|^2 = 0, & (6.43d) \end{cases}$$

where we used the relation $\psi_0(\alpha) = \frac{V}{E_\alpha - \epsilon_d}\psi_1(\alpha)$ that is valid $\forall\alpha \geq 0$, see Eq. (6.23). As before, there are N variables $\psi_0(\beta)$ and only $N-1$ variables e_α. Conditions on $\psi_0(0)$ to reduce the number of variables are determined by Eq. (6.43a), that fixes the normalization of the wave function. The condition fixing E_0 was not set before, and is now given by Eq. (6.43c). The two constraints are therefore:

$$\begin{cases} |\psi_0(0)|^2 = 1 - \sum_{\beta\geq 1} |\psi_0(\beta)|^2, \\ \sum_{\beta=0}^{N} \left(E_\beta - \epsilon_d\right)|\psi_0(\beta)|^2 = 0, \end{cases} \tag{6.44}$$

and injecting (6.43a) in (6.43c) reformulates the constraint on E_0:

$$\left(E_0 - \epsilon_d\right)\left(1 - \sum_{\beta\geq 1} |\psi_0(\beta)|^2\right) + \sum_{\beta\geq 1}\left(E_\beta - \epsilon_d\right)|\psi_0(\beta)|^2 = 0$$

$$\sum_{\beta\geq 1}\left(E_\beta - E_0\right)|\psi_0(\beta)|^2 = (\epsilon_d - E_0). \tag{6.45}$$

In the trivial limit $V = 0$, the impurity is totally decoupled from the bath, so that $\psi_0(\beta) = \delta_{\beta 0}$, and the energy of the impurity is $E_0 = \epsilon_d$, a result that is easily seen in Eq. (6.4.2.2). To express every $|\psi_0(\beta)|^2$, it is used that Eq. (6.43a) is the $\alpha = 0$ case of Eq. (6.43b), such that the latter is extended for all α:

$$\sum_{\beta\geq 0} \frac{E_\beta - \epsilon_d}{E_\beta - e_\alpha}|\psi_0(\beta)|^2 = 1, \quad \forall\alpha \geq 0. \tag{6.46}$$

An equation similar to Eq. (6.23) is found yet with a square matrix $N \times N$ including z_0 and E_0, and the system to solve is:

$$\left(E_\beta - \epsilon_d\right)z_\beta = \frac{\det(\hat{A}_\beta)}{\det(\hat{A})}, \quad \text{with} \quad z_\beta = |\psi_0(\beta)|^2. \tag{6.47}$$

In this case, the matrix $\hat{A}_\beta$ is defined by:

$$\hat{A}_\beta = \begin{pmatrix} \frac{1}{E_0 - e_0} & \dots 1 \dots & \frac{1}{E_0 - e_N} \\ \vdots & \ddots \vdots \ddots & \vdots \\ \vdots & \ddots \vdots \ddots & \vdots \\ \frac{1}{E_N - e_0} & \dots 1 \dots & \frac{1}{E_N - e_N} \end{pmatrix}. \tag{6.48}$$

As for Eq. (6.30), it is found that:

$$\det(\hat{A}) = \frac{\prod\limits_{\substack{\gamma\geq 0,\\ \gamma\neq\beta}} (E_\beta - E_\gamma)}{\prod\limits_{\gamma\geq 0} (E_\beta - e_\gamma)} \det(\hat{A}_\beta), \tag{6.49}$$

and injecting into Eq. (6.47):

$$\boxed{z_\beta = \frac{\prod\limits_{\gamma\geq 1} (E_\beta - e_\gamma)}{\prod\limits_{\substack{\gamma\geq 0,\\ \gamma\neq\beta}} (E_\beta - E_\gamma)}} \quad \text{and} \quad \boxed{\sum_{\beta\geq 1} z_\beta (E_\beta - E_0) = \sum_{\beta\geq 1} (E_\beta - e_\beta) = (\epsilon_d - E_0).} \tag{6.50}$$

With this definition, the change of $N-1$ variables e_α to N variables z_α must be considered. At the end, the constraint Eq. (6.43a) is enforced to suppress the extra degree of freedom. As before, the determinant of the Jacobian matrix associated to this change of variables is determined. In this case, it is simpler to express the derivative in the reversed way: at the end, the determinant is just the inverse of the one found here. These elements read:

$$\left(\frac{\partial z_\beta}{\partial e_\alpha}\right)_{\alpha,\beta\geq 1} = \frac{-z_\beta}{E_\beta - e_\alpha}, \tag{6.51}$$

so that its determinant is defined as:

$$\det(\hat{J}_{e\to\Psi}) = \det\left(\frac{\partial z_\beta}{\partial e_\alpha}\right)^{-1} = \left(\prod_\gamma -z_\gamma\right)^{-1} \det(\hat{C}_J)^{-1}, \tag{6.52}$$

where $\hat{C}_J$ is a Cauchy matrix whose determinant is known, so that:

$$\begin{aligned}
\det(\hat{J}_{e\to\Psi}) &= (-1)^{N-1} \left(\prod_{\beta\geq 1} z_\beta \frac{\prod\limits_{1\leq\delta<\gamma} (E_\delta - E_\gamma)(e_\gamma - e_\delta)}{\prod\limits_{\delta,\gamma} (E_\delta - e_\gamma)}\right)^{-1} \\
&= (-1)^{N-1} \left(\frac{\prod\limits_{\beta,\gamma\geq 1} (E_\beta - e_\gamma)}{\prod\limits_{\substack{\beta\geq 1,\gamma\geq 0,\\ \gamma\neq\beta}} (E_\beta - E_\gamma)} \frac{\prod\limits_{1\leq\delta<\gamma} (E_\delta - E_\gamma)(e_\gamma - e_\delta)}{\prod\limits_{\delta,\gamma} (E_\delta - e_\gamma)}\right)^{-1} \\
&= (-1)^{N-1} \left(\prod_{1\leq\delta<\gamma} \frac{(e_\gamma - e_\delta)}{(E_\delta - E_\gamma)} \prod_{\delta\geq 1} \frac{1}{(E_\delta - E_0)}\right)^{-1},
\end{aligned} \tag{6.53}$$

$$\Rightarrow \boxed{\det(\hat{J}_{e\to\psi}) = (-1)^{N-1}\left(\prod_{1\le\delta<\gamma}\frac{(E_\delta - E_\gamma)}{(e_\gamma - e_\delta)}\prod_{\delta\ge 1}(E_\delta - E_0)\right)} \tag{6.54}$$

We arrived at the desired result, the change of variables being completely defined. However, some terms can not trivially be replaced using this change of variables: the first one is in the exponential of the PDF, and involves the sum of the squared free eigenvalues. The second one is in the delta function (enforcing the normalization of the wave function), and corresponds to the sum of every r_β. Simple results are found for these two terms, and their derivation is presented now.

6.4.2.3 Sum of Squared Eigenvalues

To replace the term $\sum_{\alpha=0}^{N} e_\alpha^2$ which appears in the exponential in Eq. (6.15), the sum of the free eigenvalues squared is required. It is easily derived using the trace of the square matrix $\hat{G}$, which has the useful property not to depend on the basis:

$$\begin{aligned}
\sum_{\alpha=0}^{N} e_\alpha^2 &= \mathrm{Tr}(G^2) = \mathrm{Tr}\Big((H - V(\delta_{i1}\delta_{j0} + \delta_{i0}\delta_{j1}))^2\Big)\\
&= \sum_{\alpha=0}^{N} E_\alpha^2 + V^2\mathrm{Tr}\Big((\delta_{i1}\delta_{j0} + \delta_{i0}\delta_{j1})^2\Big) - \mathrm{Tr}\Big(2V\sum_{k=0}^{N} H_{ik}(\delta_{k1}\delta_{j0} + \delta_{k0}\delta_{j1})\Big)\\
&= \sum_{\alpha=0}^{N} E_\alpha^2 + 2V^2 - \mathrm{Tr}\Big(2V\sum_{k=0}^{N}\sum_{\alpha=0}^{N} E_\alpha\psi_i(\alpha)\psi_k(\alpha)(\delta_{k1}\delta_{j0} + \delta_{k0}\delta_{j1})\Big)\\
&= \sum_{\alpha=0}^{N} E_\alpha^2 + 2V^2 - 2V\sum_{j=i=0}^{N}\sum_{\alpha=0}^{N} E_\alpha\Big(\psi_i(\alpha)\psi_1(\alpha)\delta_{j0} + \psi_i(\alpha)\psi_0(\alpha)\delta_{j1}\Big)\\
&= \sum_{\alpha=0}^{N} E_\alpha^2 + 2V^2 - 4V\sum_{\alpha=0}^{N} E_\alpha\psi_0(\alpha)\psi_1(\alpha).
\end{aligned} \tag{6.55}$$

Using Eq. (6.23), the sum is easily expressed with the new variables:

$$\boxed{\sum_{\alpha=0}^{N} e_\alpha^2 = \sum_{\alpha=0}^{N} E_\alpha^2 - 4\sum_{\alpha=0}^{N} E_\alpha(E_\alpha - \epsilon_d)z_\alpha + 2V^2}. \tag{6.56}$$

6.4.2.4 Sum of Squared Eigenvectors

The sum of all the old eigenvectors appears in a Dirac delta to ensure the normalization of $\phi_1(\beta)$. This sum must also be expressed as a function of our new variables. Using first the property that $r_0 = 0$, it reads:

$$\sum_{\beta=0}^{N} r_\beta = \sum_{\beta=1}^{N} r_\beta = \sum_{\beta=1}^{N} \frac{1}{V^2} \frac{\prod_{\gamma\geq 1}(E_\gamma - e_\beta)}{\prod_{\substack{\gamma\geq 1,\\ \gamma\neq\beta}}(e_\gamma - e_\beta)} \left(e_\beta - E_0\right) = \frac{-1}{V^2} \sum_{\beta=1}^{N} \frac{\prod_{\gamma\geq 0}(E_\gamma - e_\beta)}{\prod_{\substack{\gamma\geq 1,\\ \gamma\neq\beta}}(e_\gamma - e_\beta)}$$

$$= \frac{-1}{V^2} \sum_{\beta=1}^{N} \frac{(-1)^N \left(-e_\beta^{N+1} + \left(\sum_{\alpha=0}^{N} E_\alpha\right) e_\beta^N - \left(\sum_{0\leq\alpha<\delta}^{N} E_\alpha E_\delta\right) e_\beta^{N-1} \right) + P(e_\beta^{N-2})}{(-1)^{N-1} \prod_{\substack{\gamma\geq 1,\\ \gamma\neq\beta}}(e_\beta - e_\gamma)}$$

$$= \frac{-1}{V^2} \left(\sum_{\beta=1}^{N} \left(\frac{e_\beta^{N+1}}{\prod_{\substack{\gamma\geq 1,\\ \gamma\neq\beta}}(e_\beta - e_\gamma)} \right) - \left(\sum_{\alpha=0}^{N} E_\alpha\right)\left(\sum_{\alpha=1}^{N} e_\alpha\right) + \left(\sum_{0\leq\alpha<\delta}^{N} E_\alpha E_\delta\right) \right), \tag{6.57}$$

where $P(e_\beta^{N-2})$ is a polynom of order $N-2$ in e_β, such that it vanishes due to Eq. (6.34). The terms concerning e_β to the power N and $N-1$ are also simplified with the same equation. However, a last term with a power $r = N+1$ is left to calculate. The set of equations Eq. (6.34) are extended using complex analysis to express this sum also in simple terms. For this purpose, the following complex valued function $f(z)$ is defined:

$$f(z) = \frac{z^{N+1}}{\prod_{i=1}^{N}(z - e_i)}. \tag{6.58}$$

Using the residue theorem, this function can be integrated over a contour C enclosing every simple pole e_i:

$$\int_C f(z)\,\mathrm{d}z = 2\pi i \sum_{j=1}^{N} \lim_{z\to e_j}(z - e_j) f(z) = 2\pi i \sum_{j=1}^{N} \frac{e_j^{N+1}}{\prod_{\substack{i=1\\ i\neq j}}^{N}(e_i - e_j)}, \tag{6.59}$$

that is, up to a factor $2\pi i$, the sum that needs to be computed. The function $f(z)$ is analytic, so that a Laurent series can be defined on the contour C that will converge to $f(z)$. This expansion reads:

$$f(z) = \frac{z^{N+1}}{\prod_{i=1}^{N}(z - e_i)} = \frac{z^{N+1}}{z^N \prod_{i=1}^{N}\left(1 - \frac{e_i}{z}\right)} = z \prod_{i=1}^{N} \frac{1}{\left(1 - \frac{e_i}{z}\right)}. \tag{6.60}$$

This function can be developed in series:

$$f(z) = z \prod_{i=1}^{N} \left(\sum_{j=0}^{+\infty} \left(\frac{e_i}{z}\right)^j \right), \tag{6.61}$$

which corresponds to the Laurent series of $f(z)$. The first orders in $\frac{1}{z}$ are developed explicitely to compute the integral, which gives:

$$\begin{aligned}
\int_{\mathcal{C}} f(z)\,\mathrm{d}z &= \int_{\mathcal{C}} z \prod_{i=1}^{N} \left(1 + \frac{e_i}{z} + \left(\frac{e_i}{z}\right)^2 + o\left(\frac{1}{z^2}\right)\right) \mathrm{d}z \\
&= \int_{\mathcal{C}} z \left(1 + \left(\sum_{i=1}^{N} e_i\right)\frac{1}{z} + \left(\sum_{1\le i<j}^{N} e_i e_j + \sum_{i=1}^{N} e_i^2\right)\frac{1}{z^2} + o\left(\frac{1}{z^2}\right)\right) \mathrm{d}z \\
&= \int_{\mathcal{C}} z + \left(\sum_{i=1}^{N} e_i\right) + \frac{1}{z}\left(\sum_{1\le i<j}^{N} e_i e_j + \sum_{i=1}^{N} e_i^2\right) + o\left(\frac{1}{z}\right) \mathrm{d}z.
\end{aligned} \tag{6.62}$$

Almost all terms in the integral can be eliminated: $o\left(\frac{1}{z}\right)$ terms will tend to 0 when the radial part of the contour $\mathcal{C}$ will be set to $+\infty$, since they are all fractions of the type $\frac{1}{z^\alpha}$ with $\alpha > 1$. The constant and linear terms will simply be 0 using Cauchy integral theorem, since they have no poles. Only the z^{-1} term is left to calculate, which can be regularized to properly use the residue theorem around the pole at $z = 0$:

$$\begin{aligned}
\int_{\mathcal{C}} f(z)\,\mathrm{d}z &= \left(\sum_{1\le i<j}^{N} e_i e_j + \sum_{i=1}^{N} e_i^2\right) \lim_{\epsilon\to 0^+} \int_{\mathcal{C}} \frac{1}{z} e^{-i\epsilon z} \mathrm{d}z \\
&= \left(\sum_{1\le i<j}^{N} e_i e_j + \sum_{i=1}^{N} e_i^2\right) \lim_{\epsilon\to 0^+} 2\pi i\ \mathrm{sgn}(\epsilon) \\
&= 2\pi i \left(\sum_{1\le i<j}^{N} e_i e_j + \sum_{i=1}^{N} e_i^2\right).
\end{aligned} \tag{6.63}$$

Putting this last equation in Eq. (6.59), the desired result reads:

$$\sum_{j=1}^{N} \frac{e_j^{N+1}}{\prod_{\substack{i=1\\i\neq j}}^{N}(e_i - e_j)} = \left(\sum_{1\leq i<j}^{N} e_i e_j + \sum_{i=1}^{N} e_i^2\right). \tag{6.64}$$

The sum of eigenvectors is thus given by:

$$\sum_{\beta=0}^{N} r_\beta = \frac{-1}{V^2}\left(\sum_{i=1}^{N} e_i^2 + \sum_{1\leq i<j}^{N} e_i e_j - \left(\sum_{\alpha=0}^{N} E_\alpha\right)\left(\sum_{\alpha=1}^{N} e_\alpha\right) + \sum_{0\leq\alpha<\delta}^{N} E_\alpha E_\delta\right). \tag{6.65}$$

Further simplifications can be done with some rearrangements in the previous equation:

$$\left(\sum_{i=1}^{N} x_i\right)^2 = \sum_{i=1}^{N} x_i \sum_{j=1}^{N} x_j = \sum_{i=1}^{N} x_i^2 + \sum_{\substack{i=1\\i\neq j}}^{N} x_i x_j = \sum_{i=1}^{N} x_i^2 + 2\sum_{1\leq i<j}^{N} x_i x_j$$
$$\Leftrightarrow \sum_{1\leq i<j}^{N} x_i x_j = \frac{1}{2}\left(\left(\sum_{i=1}^{N} x_i\right)^2 - \sum_{i=1}^{N} x_i^2\right). \tag{6.66}$$

This representation is also used for the sums including $E_i E_j$ and $e_i e_j$, which leads to the following equation:

$$\sum_{\beta=0}^{N} r_\beta = \frac{-1}{V^2}\left(\frac{1}{2}\sum_{i=1}^{N} e_i^2 + \frac{1}{2}\left(\sum_{i=1}^{N} e_i\right)^2 - \left(\sum_{\alpha=0}^{N} E_\alpha\right)\left(\sum_{\alpha=1}^{N} e_\alpha\right) + \frac{1}{2}\left(\sum_{i=0}^{N} E_i\right)^2 - \frac{1}{2}\sum_{i=0}^{N} E_i^2\right)$$
$$= \frac{-1}{2V^2}\left(\left(\sum_{i=1}^{N} e_i^2 - \sum_{i=0}^{N} E_i^2\right) + \left(\sum_{i=1}^{N} e_i - \sum_{i=0}^{N} E_i\right)^2\right). \tag{6.67}$$

For the first part, Eq. (6.56) is used, and for the second one, it can be derived in the same spirit than in the previous Sect. 6.4.2.3:

$$\sum_{i=0}^{N} e_\alpha = \mathrm{Tr}(G) = \mathrm{Tr}\left(H - V(\delta_{i0}\delta_{j1} + \delta_{i1}\delta_{j0})\right) = \mathrm{Tr}(H) = \sum_{i=0}^{N} E_\alpha. \tag{6.68}$$

The final result is hence given by:

$$\sum_{\beta=0}^{N} r_\beta = \frac{-1}{2V^2}\left(\sum_{\alpha=0}^{N} E_\alpha^2 - 4\sum_{\alpha=0}^{N} E_\alpha(E_\alpha - \epsilon_d)z_\alpha + 2V^2 - \epsilon_d^2 - \sum_{i=0}^{N} E_i^2 + \epsilon_d^2\right)$$

$$= \frac{-1}{2V^2}\left(-4\sum_{\alpha=0}^{N} E_\alpha(E_\alpha - \epsilon_d)z_\alpha\right) - 1$$

$$\Rightarrow \boxed{\sum_{\beta=0}^{N} r_\beta = \frac{2}{V^2}\sum_{\alpha=0}^{N} E_\alpha(E_\alpha - \epsilon_d)z_\alpha - 1}. \tag{6.69}$$

6.4.3 Distribution

Every part of the change of variables being defined, the joint distribution of eigenvalues and eigenvectors of the random impurity matrix H can be derived:

$$P\Big(\{E_\alpha\}, \{z_\alpha\}\Big) = P\Big(\{e_\alpha\}, \{r_\alpha\}\Big)|\det(\hat{J}_{e\to\Psi})||\det(\hat{J}_{\phi\to E})|. \tag{6.70}$$

The first change of variables reads:

$$P\Big(\{e_\alpha\}, \{E_\alpha\}\Big) = P\Big(\{e_\alpha\}, \{r_\alpha\}\Big)|\det(\hat{\hat{J}}_{\phi\to E})|$$

$$\propto \prod_{1\le\alpha<\beta}^{N}(e_\alpha - e_\beta)\frac{V\prod_{\substack{\gamma\ge1,\\ \gamma\ne\alpha}}(e_\gamma - e_\alpha)^{\frac{1}{2}}}{(e_\alpha - E_0)^{\frac{1}{2}}\prod_{\gamma\ge1}(E_\gamma - e_\alpha)^{\frac{1}{2}})}\prod_{1\le\gamma<\delta}\frac{(E_\delta - E_\gamma)}{(e_\gamma - e_\delta)}$$

$$\times\prod_{\gamma\ge1}\Big(E_\gamma - E_0\Big)\,\delta\left(\sum_{\gamma\ge1}^{N} r_\gamma - 1\right)e^{-\frac{1}{4\sigma}\sum_{\gamma\ge1}^{N} e_\gamma^2}$$

$$\propto \frac{\prod_{1\le\gamma<\delta}(E_\delta - E_\gamma)(e_\delta - e_\gamma)}{\prod_{\delta,\gamma\ge1}(E_\gamma - e_\delta)^{\frac{1}{2}}}\prod_{\gamma\ge1}\frac{\Big(E_\gamma - E_0\Big)}{\Big(e_\gamma - E_0\Big)^{\frac{1}{2}}}\delta\left(\sum_{\gamma\ge1}^{N} r_\gamma - 1\right)e^{-\frac{1}{4\sigma}\sum_{\gamma\ge1}^{N} e_\gamma^2}. \tag{6.71}$$

Then the second one:

$$P\Big(\{E_\alpha\},\{z_\alpha\}\Big) = P\Big(\{e_\alpha\},\{E_\alpha\}\Big)\delta\left(z_\alpha - \frac{\prod\limits_{\gamma\geq 1}(E_\alpha - e_\gamma)}{\prod\limits_{\substack{\gamma\geq 0,\\ \gamma\neq\alpha}}(E_\alpha - E_\gamma)}\right)|\det(\hat{J}_{e\to\Psi})|$$

$$\propto \frac{\prod\limits_{1\leq\gamma<\delta}(E_\delta - E_\gamma)(e_\delta - e_\gamma)}{\prod\limits_{\delta,\gamma\geq 1}(E_\gamma - e_\delta)^{\frac{1}{2}}}\prod_{\gamma\geq 1}\frac{\big(E_\gamma - E_0\big)}{\big(e_\gamma - E_0\big)^{\frac{1}{2}}}\delta\left(\sum_{\gamma\geq 1}^{N} r_\gamma - 1\right)$$

$$\times\,\delta\left(z_\alpha - \frac{\prod\limits_{\gamma\geq 1}(E_\alpha - e_\gamma)}{\prod\limits_{\substack{\gamma\geq 0,\\ \gamma\neq\alpha}}(E_\alpha - E_\gamma)}\right)\left(\prod_{1\leq\delta<\gamma}\frac{(E_\delta - E_\gamma)}{(e_\gamma - e_\delta)}\prod_{\delta\geq 1}(E_\delta - E_0)\right)e^{-\frac{1}{4\sigma}\sum\limits_{\gamma\geq 1}^{N}e_\gamma^2} \tag{6.72}$$

$$\propto \frac{\prod\limits_{1\leq\gamma<\delta}(E_\delta - E_\gamma)^2}{\prod\limits_{\delta,\gamma\geq 1}(E_\gamma - e_\delta)^{\frac{1}{2}}}\prod_{\gamma\geq 1}\frac{\big(E_\gamma - E_0\big)^2}{\big(e_\gamma - E_0\big)^{\frac{1}{2}}}\delta\left(\sum_{\gamma\geq 1}^{N} r_\gamma - 1\right)e^{-\frac{1}{4\sigma}\sum\limits_{\gamma\geq 1}^{N}e_\gamma^2}, \tag{6.73}$$

where $\prod\limits_{\gamma,\delta\geq 1}(E_\gamma - e_\delta) = \prod\limits_{\gamma\geq 1}(z_\gamma)\prod\limits_{\substack{\gamma\geq 1,\delta\geq 0,\\ \delta\neq\gamma}}(E_\gamma - E_\delta) = \prod\limits_{\gamma\geq 1}(z_\gamma)\prod\limits_{\gamma\geq 1}(E_\gamma - E_0)\prod\limits_{1\leq\gamma<\delta}$ $(E_\gamma - E_\delta)^2$, and using Eqs. (6.69) and (6.56) for the delta and the exponential:

$$P\Big(\{E_\alpha\},\{z_\alpha\}\Big) \propto \prod_{1\leq\gamma<\delta}(E_\delta - E_\gamma)\prod_{\gamma\geq 1}\frac{\big(E_\gamma - E_0\big)^{3/2}}{\big(e_\gamma - E_0\big)^{\frac{1}{2}}}e^{-\frac{1}{4\sigma}\sum\limits_{\gamma=0}^{N}E_\gamma^2}$$

$$\times\prod_{1\leq\alpha}\frac{1}{(z_\alpha)^{\frac{1}{2}}}\delta\left(\sum_{\gamma=0}^{N}E_\gamma(E_\gamma - \epsilon_d)z_\gamma - V^2\right)e^{\frac{1}{\sigma}\sum\limits_{\gamma=0}^{N}E_\gamma(E_\gamma-\epsilon_d)z_\gamma}e^{-\frac{V^2}{2\sigma}} \tag{6.74}$$

Eigenvalues of the free problem are still present, and they are replaced by identifying $(z_0)^{\frac{1}{2}} = \left(\frac{\prod_{\gamma\geq 1}(E_0-e_\gamma)}{\prod_{\gamma\geq 1}(E_0-E_\gamma)}\right)^{\frac{1}{2}}$. The delta function is applied to cancel the exponential term:

$$P\Big(\{E_\alpha\},\{z_\alpha\}\Big) \propto \prod_{0\leq\gamma<\delta}(E_\delta - E_\gamma)\prod_{\alpha\geq 0}\frac{1}{(z_\alpha)^{\frac{1}{2}}}\delta\left(\sum_{\gamma=0}^{N}E_\gamma(E_\gamma - \epsilon_d)z_\gamma - V^2\right)e^{-\frac{1}{4\sigma}\sum\limits_{\gamma=0}^{N}E_\gamma^2}. \tag{6.75}$$

However, this equations still contains the two additional variables, and the constraints $z_0 = 1 - \sum\limits_{\gamma\geq 1} z_\gamma$ and $(E_0 - \epsilon_d) = \sum\limits_{\gamma\geq 1}(E_0 - E_\gamma)z_\gamma$ must be introduced. The latter can be simplified $E_0(1 - \sum\limits_{\gamma\geq 1} z_\gamma) - \epsilon_d = -\sum\limits_{\gamma\geq 1} E_\gamma z_\gamma$, and injecting the first constraint leads to $\sum_{\gamma\geq 0} E_\gamma z_\gamma = \epsilon_d$. Gathering everything, the distribution of the perturbed problem can be finally written:

$$P_{\text{pert}}\Big(\{E_\alpha\},\{z_\alpha\}\Big) \propto \prod_{0\leq\gamma<\delta} |E_\delta - E_\gamma| e^{-\frac{1}{2\sigma^2}\sum_{\gamma=0}^{N} E_\gamma^2} \prod_{\alpha\geq 0} \frac{1}{\sqrt{z_\alpha}} \delta\left(V^2 - \sum_{\gamma=0}^{N} E_\gamma (E_\gamma - \epsilon_d) z_\gamma\right)$$
$$\times\, \delta\left(\epsilon_d - \sum_{\gamma=0}^{N} E_\gamma z_\gamma\right) \delta\left(1 - \sum_{\gamma=0}^{N} z_\gamma\right). \tag{6.76}$$

This equation is the main result of this chapter and is exact (up to a norm) for any parameters or system size N. These three constraints have been verified numerically, at machine precision for each realization of the random matrix which ensures the validity of the presented derivation. The first one, $V^2 = \sum_{\gamma=0}^{N} E_\gamma (E_\gamma - \epsilon_d) z_\gamma$ corresponds to the constraint in Eq. (6.15) expressed with the new variables, and the two others are imposed to reduce the number of $(2N)^2$ variables in the new basis to $(2N-1)^2$ random variables that are present in the original problem.

The main difficulty that appears at this point is to treat correctly the different constraints, that are usually treated at a mean field level for large N. In the upcoming section, actual calculations with this distribution function are presented, assuming that N is large enough. Additional calculations for simpler PDF are also presented in Appendix E to gain intuition about the large-N methods employed below.

6.5 Large N Calculations of Distribution Functions

To perform actual calculations with this distribution function, some rearrangements are required. In Appendix E, large-N calculations are presented in simpler models (no perturbation, and rank-one perturbation) to gain intuition about the method, and analytical results are simpler to find in these models. The calculations are similar in the present case, but the form of the constraints (that are quadratic in E) complicates the analytical integration over random variables. In this section, the calculations are presented step by step.

6.5.1 *Simplifications for $P(n_d)$*

The delta distributions are the first elements to be reexpressed, and they are replaced by their definition through the Fourier transform as:

$$\delta\left(V^2-\sum_{\gamma=0}^{N} E_\gamma(E_\gamma-\epsilon_d)z_\gamma\right)=\int_{-\infty i}^{\infty i}\frac{\mathrm{d}\lambda}{2\pi}\exp\left[-\lambda\left(V^2-\sum_{\gamma=0}^{N} E_\gamma(E_\gamma-\epsilon_d)z_\gamma\right)\right], \tag{6.77}$$

where λ is a Lagrange multiplier which is integrated, and the imaginary element i in the usual definition can be absorbed in λ for convenience. Transforming each constraint with their Fourier transform, Eq. (6.76) can be written:

$$\begin{aligned}P_{\text{pert}}\left(\{E_\alpha\},\{z_\alpha\}\right)\propto\prod_{\alpha\geq 0}\frac{1}{\sqrt{z_\alpha}}\int_{-\infty i}^{+\infty i}\frac{\mathrm{d}\lambda}{2\pi}\int_{-\infty i}^{+\infty i}\frac{\mathrm{d}\nu}{2\pi}\int_{-\infty i}^{+\infty i}\frac{\mathrm{d}\rho}{2\pi}\exp\Big[\sum_{\gamma=0}^{N}-\frac{1}{2\sigma^2}E_\gamma^2+\sum_{\delta<\gamma}^{N}\ln\left(|E_\delta-E_\gamma|\right)\\-\lambda\left(V^2-\sum_{\gamma=0}^{N}E_\gamma^2 z_\gamma\right)-\nu\left(-\sum_{\gamma=0}^{N}E_\gamma z_\gamma\right)-\rho\left(1-\sum_{\gamma=0}^{N}z_\gamma\right)\Big].\end{aligned} \tag{6.78}$$

This can be done provided that every term is of the same order in N, which is true in the diluted regime. This form is already more appealing to integrate over the random variables. Other simplifications will appear when integrating over these variables, in order to compute $P(n_d)$:

$$P(n_d)=\int_{-\infty}^{+\infty}\mathcal{D}E_\alpha\int_{0}^{+\infty}\mathcal{D}z_\alpha\, P_{\text{pert}}(E_\alpha,z_\alpha)\,\delta\left(n_d-\sum_{\gamma=0}^{N}\Theta(-E_\alpha)z_\alpha\right), \tag{6.79}$$

where $\mathcal{D}E_\alpha=\prod_{\alpha=0}^{N}\mathrm{d}E_\alpha$ is written for readability. Inserting the new constraint in the exponential, every term proportional to z_α can be factorized, and the integration over z_α becomes a Gaussian integration with the change of variables $\sqrt{z_\alpha}\rightarrow z'_\alpha$. The equation is thus further simplified to:

$$\begin{aligned}P(n_d)=\int_{-\infty}^{+\infty}\mathcal{D}E_\alpha\iiiint_{-\infty i}^{+\infty i}\frac{\mathrm{d}\lambda\,\mathrm{d}\nu\,\mathrm{d}\rho\,\mathrm{d}\mu}{(2\pi)^{7/2}}\exp\Big[\sum_{\gamma=0}^{N}-\frac{1}{2\sigma^2}E_\gamma^2+\sum_{0\leq\delta<\gamma}\ln\left(|E_\delta-E_\gamma|\right)\\-\frac{1}{2}\sum_{\gamma=0}^{N}\ln\left(|\lambda E_\gamma^2+\nu E_\gamma+\rho+\mu\Theta(-E_\alpha)|\right)\Big].\end{aligned} \tag{6.80}$$

The Heaviside function is not analytic and may be problematic in the following. With the knowledge of the final spectra thanks to numerical simulations, we assume that half of the eigenvalues are negative and half are positive. Such an approximation is tested for the free case in Appendix E and proved to be accurate. In the limit of large N, the steepest descend method is used to calculate the integral over all eigenvalues, corresponding to a mean field treatment of the bath if Green's functions are used. This steepest descent method follows the method presented in Ref. [25], in which the renormalization of the bath is computed for a ϕ^4 bulk interaction. The scaling of each terms have to be carefuly accounted: $E_\alpha\propto\sqrt{N}$ and $V\propto\sqrt{N}$, such that

$\lambda \propto 1$, $\nu \propto \sqrt{N}$ and ρ, $\mu \propto N$. With these scalings, the saddle-point to solve for the steepest descent reads:

$$\partial_{E_\alpha} P(n_d) = 0 \Rightarrow \sum_{\substack{\beta \geq 0, \\ \beta \neq \alpha}} \frac{1}{E_\alpha - E_\beta} = \frac{E_\alpha}{2\sigma^2} + \frac{1}{2} \frac{\lambda E_\alpha + \nu}{\lambda E_\alpha^2 + \nu E_\alpha + \rho}, \tag{6.81}$$

and by writing explicitely the scalings $\tilde{E}_\alpha \sqrt{N} = E_\alpha$, and by remarking that the sum in the left-hand-side (LHS) concerns N terms:

$$\sum_{\substack{\beta \geq 0, \\ \beta \neq \alpha}} \frac{1}{\tilde{E}_\alpha - \tilde{E}_\beta} = \frac{\tilde{E}_\alpha}{2\sigma^2} + \frac{1}{2N} \frac{\tilde{\lambda} \tilde{E}_\alpha + \tilde{\nu}}{\tilde{\lambda} \tilde{E}_\alpha^2 + \tilde{\nu} \tilde{E}_\alpha + \tilde{\rho}}, \tag{6.82}$$

and the term proportional to $1/N$ in the RHS can be safely neglected in this situation when N is large. Going to the continuous limit $E_\alpha \to E(\alpha/N)$, and $1/N \sum_\alpha^N \to \int_0^1 \mathrm{d}x$:

$$\fint_0^1 \frac{\mathrm{d}y}{E(x) - E(y)} = \frac{1}{2\sigma^2} E(x), \tag{6.83}$$

where $\fint$
represents the principal part of the integral. At this point, the density of states $u(E) = \mathrm{d}x/\mathrm{d}E$ is introduced, normalized on an interval $[-2a, 2a]$ which gives:

$$\fint_{-2a}^{2a} \mathrm{d}F \frac{u(F)}{E - F} = \frac{1}{2\sigma^2} E, \quad \forall |E| \leq 2a. \tag{6.84}$$

Following Ref. [25], it is trivial to find that in this case the density of states is the semi-circle in the support $[-\sqrt{2N}\sigma, \sqrt{2N}\sigma]$, $u(E) = \sqrt{2N\sigma^2 - E^2}/2\pi N$. Integrating over E in Eq. (6.84), it allows to introduce the density of states as the mean field solution (or order 0 solution) of the integral over $\mathcal{D}E_\alpha$ in the definition of $P(n_d)$ as:

$$P(n_d) = \iiiint_{-\infty i}^{+\infty i} \frac{\mathrm{d}\lambda\, \mathrm{d}\nu\, \mathrm{d}\rho\, \mathrm{d}\mu}{(2\pi)^{7/2}} \exp\Bigg[\int_{-\sqrt{2N}\sigma}^{0} \mathrm{d}E\, u(E) \left(\ln(|E|) - \frac{E^2}{2\sigma^2} - \frac{1}{2} \ln\left(\lambda E^2 + \nu E + \rho + \mu\right) \right)$$
$$+ \int_0^{\sqrt{2N}\sigma} \mathrm{d}E\, u(E) \left(\ln(|E|) - \frac{E^2}{2\sigma^2} - \frac{1}{2} \ln\left(\lambda E^2 + \nu E + \rho\right) \right) + \lambda V^2 + \rho + \mu n_d \Bigg]. \tag{6.85}$$

The integration over the Lagrange multiplier is also treated by steepest descent, with the saddles defining the following system of equations to solve:

$$\begin{cases} \partial_\lambda P(n_d) = 0 \Rightarrow \int\limits_{-\sqrt{2N}\sigma}^{0} \mathrm{d}E \frac{u(E)E^2}{\lambda E^2+\nu E+\rho+\mu} + \int\limits_{0}^{\sqrt{2N}\sigma} \mathrm{d}E \frac{u(E)E^2}{\lambda E^2+\nu E+\rho} = V^2, \\ \partial_\nu P(n_d) = 0 \Rightarrow \int\limits_{-\sqrt{2N}\sigma}^{0} \mathrm{d}E \frac{u(E)E}{\lambda E^2+\nu E+\rho+\mu} + \int\limits_{0}^{\sqrt{2N}\sigma} \mathrm{d}E \frac{u(E)E}{\lambda E^2+\nu E+\rho} = 0, \\ \partial_\rho P(n_d) = 0 \Rightarrow \int\limits_{-\sqrt{2N}\sigma}^{0} \mathrm{d}E \frac{u(E)}{\lambda E^2+\nu E+\rho+\mu} + \int\limits_{0}^{\sqrt{2N}\sigma} \mathrm{d}E \frac{u(E)}{\lambda E^2+\nu E+\rho} = 1, \\ \partial_\mu P(n_d) = 0 \Rightarrow \int\limits_{-\sqrt{2N}\sigma}^{0} \mathrm{d}E \frac{u(E)}{\lambda E^2+\nu E+\rho+\mu} = n_d, \end{cases}$$

$$\Rightarrow \begin{cases} \lambda V^2 + \rho + \mu n_d = 1, \\ \int\limits_{0}^{\sqrt{2N}\sigma} \mathrm{d}E \frac{u(E)}{\lambda E^2+\nu E+\rho} = 1 - n_d, \\ \int\limits_{-\sqrt{2N}\sigma}^{0} \mathrm{d}E \frac{u(E)}{\lambda E^2+\nu E+\rho+\mu} = n_d, \\ \int\limits_{-\sqrt{2N}\sigma}^{0} \mathrm{d}E \frac{u(E)E}{\lambda E^2+\nu E+\rho+\mu} + \int\limits_{0}^{\sqrt{2N}\sigma} \mathrm{d}E \frac{u(E)E}{\lambda E^2+\nu E+\rho} = 0. \end{cases} \tag{6.86}$$

Solving these equations and reinjecting the solutions in Eq. (6.85) would give the solution of this problem. However, this system has no simple analytical form in this case, in contrast to the simpler toy model of Sect. 6.3.

6.5.2 Remarks on the Weak Coupling Regime

General remarks on calculations in the weak coupling regime are discussed here, as no results have been found so far in this case. For any distribution function in the weak coupling regime, some of the previous assumptions do not hold anymore. In this case, V is subdominant and large N expansions can not be done as before. For the bath density of states, the terms in $1/N$ in Eq. (6.82) must be considered, which brings an explicit dependence of the bath density of states on the Lagrange multipliers. Nevertheless, previous results for the DOS presented in Eq. (6.4) can be used at 0th order. As the impurity is mostly decoupled from the bath, a single z_0 is larger than the others and must be taken out of the steepest descent equations, and must be treated apart. Things complicates as V/σ increases, as in the crossover more and more z_α becomes of order 1, and simplifies when they all become of order $1/N$. The bound states regime, in which calculations are tractable, is now discussed.

6.5.3 $P(E_0, z_0)$ in the Bound States Regime

An interesting quantity to look at is the distribution of the energy and of the amplitude of the outliers when $V/\sigma > \sqrt{2N}$, that is not accurately reproduced by the toy model. In this regime, two elements z_{o1} and z_{o2} are of order one, and must be taken out in Eq. (6.77), and in the following. It is also chosen that $\sigma = \sqrt{2}$. To simplify the equations, it is assumed that the two outliers are symmetric, such that:

$$E_{o1} = -E_{o2} = E_0, \quad \text{and} \quad z_{o1} = z_{o2} = z_0. \tag{6.87}$$

Due to this simplification, the multiplier ν, that holds information concerning the asymmetry of the distribution, is taken to be 0 such that the last equation of Eq. (6.86) is always true. The distribution is built by integrating Eq. (6.78) over all but two pairs of (E_0, z_0) and reads:

$$\begin{aligned} P(E_0, z_0) \propto \frac{1}{\sqrt{z_0}} \iint_{-\infty i}^{+\infty i} \mathrm{d}\lambda\, \mathrm{d}\rho \exp\Bigg[-\frac{E_0^2}{2} + \int_{-2\sqrt{N}}^{2\sqrt{N}} \mathrm{d}E\, u(E) \Big(\ln\left(|E_0^2 - E^2|\right) \\ - \frac{1}{2} \ln\left(\lambda E^2 + \rho\right) \Big) + \lambda\left(V^2 - 2E_0^2 z_0\right) + \rho\left(1 - 2z_0\right) \Bigg]. \end{aligned} \tag{6.88}$$

In this situation, deriving a similar set of equations as Eq. (6.86), a closed form is found for the saddle-point:

$$\begin{cases} \lambda = \frac{N}{2(V^2 - 2E_0^2 z_0)} - \frac{1}{2(1-2z_0)}, \\ \rho = \frac{V^2 - 2E_0^2 z_0}{2(1-2z_0)}. \end{cases} \tag{6.89}$$

Then, the solutions to the saddle-point equations are inserted in Eq. (6.88), and numerical integration is used for the integrals over the energy in the exponential. For $V/\sigma = 2\sqrt{2N}$, the distribution is plotted in the right panel of Fig. 6.7. In the left panel, the distribution of the energy and the amplitude of one of the outliers is plotted, and the distribution found to be wider than the one computed through Eq. (6.88). This effect is due to the assumption that the two outliers are symmetric in energy and in amplitude, which is true only in average: in the middle panel of the figure, only realizations such that $||E_{o1}| - |E_{o2}|| < \epsilon$ (with $\epsilon = 1e-2$ in this case) are considered, and the distribution is closer to the computed one, and would be the same for $\epsilon \to 0$. This confirms the validity of the large N analysis.

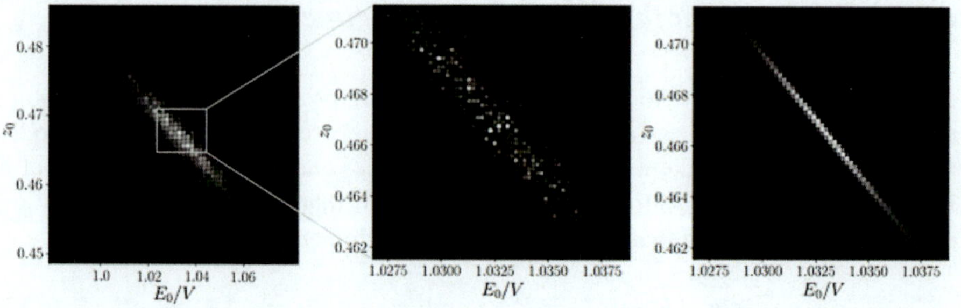

Fig. 6.7 Joint distribution of the amplitude and of the energy of the positive outlier in the bound states regime. The left panel shows the result of the numerical sampling and the right one the result of the large N calculation. The middle panel corresponds to the numerical sampling with a condition of symmetry on the outliers, to fit the assumptions taken in the analytical calculation

References

1. Wigner EP (1951) On the statistical distribution of the widths and spacings of nuclear resonance levels. Math Proc Cambridge Philos Soc 47:790–798
2. Porter CE, Thomas RG (1956) Fluctuations of nuclear reaction widths. en. Phys Rev 104:483–491. ISSN: 0031-899X
3. Mehta ML (1960) On the statistical properties of the level-spacings in nuclear spectra. en. Nucl Phys 18:395–419. ISSN: 0029-5582
4. Dyson FJ (1962) Statistical theory of the energy levels of complex systems. I. J Math Phys 3:140–156. ISSN: 0022-2488
5. Wigner EP (1955) Characteristic vectors of bordered matrices with infinite dimensions. Ann Math 62:548–564. ISSN: 0003-486X
6. Mehta ML, Gaudin M (1960) On the density of Eigenvalues of a random matrix. en. Nucl Phys 18:420–427. ISSN: 0029-5582
7. Pastur LA (1972) On the spectrum of random matrices. en. Theor Math Phys 10:67–74. ISSN: 1573-9333
8. Luo F, Zhong J, Yang Y, Scheuermann RH, Zhou J (2006) Application of random matrix theory to biological networks. en. Phys Lett A 357:420–423. ISSN: 0375-9601
9. Bouchaud JP, Potters M (2009) Financial applications of random matrix theory: a short review. arXiv:0910.1205
10. Mehta ML (2004) Random matrices 3rd. Academic Press, New York
11. Beenakker CWJ (1997) Random-matrix theory of quantum transport. Rev Mod Phys 69:731–808
12. Potters M, Bouchaud J-P (2020) A first course in random matrix theory: for physicists, engineers and data scientists. Cambridge University Press
13. Kaul RK, Ullmo D, Baranger HU (2003) Mesoscopic fluctuations in quantum dots in the Kondo regime. en. Phys Rev B 68:161305. ISSN: 0163-1829, 1095-3795 (2003)
14. Kettemann S, Mucciolo ER (2006) Free magnetic moments in disordered metals. en. J Exp Theor Phys Lett 83:240–245. ISSN: 0021-3640, 1090-6487
15. Kettemann S, Mucciolo ER (2007) Disorder-quenched Kondo effect in mesoscopic electronic systems. en. Phys Rev B 75:184407. ISSN: 1098-0121, 1550-235X
16. Liu DE, Burdin S, Baranger HU, Ullmo D (2012) Mesoscopic Anderson box: connecting weak to strong coupling. Phys Rev B 85:155455
17. Elattari B, Kagalovsky V, Weidenmüller HA (1996) Chaotic scattering with resonance enhancement. en. Nucl Phys A 606:86–94. ISSN: 0375-9474

18. Elattari B, Kagalovsky V, Weidenmüller HA (1998) Effect of resonances on diffusive scattering. Phys Rev B 57:11258–11264
19. Efetov K (1983) Supersymmetry and theory of disordered metals. Adv Phys 32:53–127. ISSN: 0001-8732
20. Verbaarschot JJM, Weidenmüller HA, Zirnbauer MR (1985) Grassmann integration in stochastic quantum physics: the case of compound-nucleus scattering. en. Phys Rep 129:367–438. ISSN: 0370-1573
21. Verbaarschot JJM, Zirnbauer MR (1984) Replica variables, loop expansion, and spectral rigidity of random-matrix ensembles. en. Ann Phys 158:78–119. ISSN: 0003-4916
22. Aleiner IL, Matveev KA (1998) Shifts of random energy levels by a local perturbation. en. Phys Rev Lett 80:814–816. ISSN: 0031-9007, 1079-7114
23. Bogomolny E (2017) Modification of the Porter-Thomas distribution by rank-one interaction. en. Phys Rev Lett 118:022501. ISSN: 0031-9007, 1079-7114
24. Knuth DE (1997) The art of computer programming, volume 1, 3rd ed. Fundamental algorithms. Addison Wesley Longman Publishing Co, Inc, USA. ISBN: 0201896834
25. Brézin E, Itzykson C, Parisi G, Zuber JB (1978) Planar diagrams. en. Commun Math Phys 59:35–51. ISSN: 1432-0916
26. Léger SEA (2019) Observation of quantum many-body effects due to zero point fluctuations in superconducting circuits. en. Nat Commun 391

Part IV
Conclusions and Perspectives

Chapter 7
Conclusion and Perspectives

In this work, correlations in quantum impurity problems have been investigated from the perspective of natural orbitals (NOs), a concept that has been borrowed from the field of quantum chemistry. Quantum impurity problems, within which the Coulomb interaction between electrons occupying the impurity induces strong correlations between particles, exhibit some drastic simplifications when they are expressed through natural orbitals. This representation reveals the hierarchical structure of correlations in these system, in which only a local interaction acts within a large bath of non-interacting particles, compared to fully interacting lattice systems whose correlations are experienced equally by all degrees of freedom.

The hierarchical structure of correlations was firstly studied with the non-perturbative numerical renormalization group algorithm in different regimes of the parameter space of two quantum impurity models where particles experience different degrees of correlations. An exponential decay of correlations that was predicted theoretically was observed in all regimes, with a slower decay rate when correlations are increasing. An ansatz for the wave function was proposed, based on a separation of natural orbitals in a correlated and uncorrelated spaces according to their occupancy in a given state.

Then, an algorithm tailored to exploit this hierarchical structure of correlations was developed, which succeeded in simulating ground-state properties at equilibrium of quantum impurity systems in a polynomial complexity with the system size N, making possible the simulation of impurity surrounded by large baths containing up to tens of thousands of particles, with a controllable accuracy. The efficiency of the algorithm was demonstrated with the computation of the large correlation cloud surrounding the impurity, reflecting the screening of the local degeneracy by electrons of the bath over the characteristic scale T_K. This quantity was also computed in a two dimensional bath in a strongly correlated regime, which was not possible with previous numerical methods on such large spatial scales.

M. Debertolis, *Quantum Impurity Problems in the Framework of Natural Orbitals*, Springer Theses, https://doi.org/10.1007/978-3-031-47233-6_7

Taking advantage of the efficiency of the algorithm, a statistical study of correlations spreading in a disordered host around a quantum impurity has been performed. Distribution functions of static quantities in the ground state have been computed, having access to tens of thousands of disorder realizations. The study focused especially on the distribution of Kondo temperatures, the relevant scales related to most of the ground state properties. Charge and spin disorder were investigated, and in both case the distribution shown to be multimodal, due to the competition between different mechanisms.

The study concerning charge disorder in the SIAM confirmed the existence of unscreened impurities at low temperatures, predicted by previous studies to explain non-Fermi liquid behavior in real setups in presence of disorder. Additional features related to charge fluctuations on the impurity were discovered in this systems when disorder is not too large. However, correlations in the disordered SIAM must be further investigated in the strong disordered regime (where the localization length is of the order of the Kondo length), since the universal relation between the local T_K and the true extent of spatial correlations remains unclear in these strongly disordered systems.

Finally, the properties of these disordered systems were studied at a mesoscopic level through a random matrix model of quantum impurity, in which microscopic information concerning the bath are lost. Results from the random matrix theory has been used to build the joint distribution function of energies and eigenstates of this problem, with which any local property concerning the impurity can be computed in average. A toy model and large N calculations were then presented in the different regimes of the model, bringing analytical formulations for certain distribution functions.

Going further with the NO idea requires various extensions of its current implementation. With the computation of dynamical quantities, the NO method could become an efficient impurity solver for multi-orbital DMFT calculations in highly correlated systems that currently lacks accurate description for low temperatures and real frequencies data. The first step will be to implement dynamics, for example using Krylov spaces, or methods based on the time-dependent variational principle, to approximate the resolvent and thus the frequency-dependent Green's function. These functions are needed to perform DMFT studies, but are also relevant on their own to compute transport properties of single impurity problems in quantum dots.

To tackle problems with even more complex internal structure (e.g. with orbital degrees of freedom), we could think of hybrid methods involving NO and DMRG, where the DMRG would describe the correlated part of impurities, going beyond exact diagonalization and allowing to deal with larger correlated spaces. On top of the MPS representation of the correlated sector, the NO algorithm will handle the environment to give the complete picture of these complex systems. These extensions could make it possible to study complicated composite problems such as multiorbital Hubbard models, via the simulation of the underlying interacting multiorbital impurity problems.

The NO method could be also useful to answer other open questions concerning the physics of quantum impurity problems. The description of microscopic properties at large scale could be used to tackle the RKKY coupling of several Kondo impurities in different configurations, or any small cluster of impurities coupled to large environments. In a similar context, the NO ansatz could be extended to bosonic systems, in order to simulate precise setups including interfaces, circuit lithography etc. stimulated by the great advances in superconducting qubits [1].

In correlated systems, computing excited states is also a challenge, especially for the study of the eigenstate thermalization hypothesis with macroscopic reservoirs, in which a finite size scaling could be done over few decades in the system size, instead of exact diagonalization studies that are limited to one order of magnitude. Excited states could be found in practice with usual ways such as shift-inverse method to target special energy windows, provided the NO ansatz stays valid.

Finally, in a wider framework, the representation of quantum impurity problems with NOs opens the door to a first-principles study of Kondo correlations, since ab-initio methods rely on the same type of ansatz that we used here. Embedding ab-initio descriptions in this context would be an ambitious project for the future.

Reference

1. Léger SEA (2019) Observation of quantum many-body effects due to zero point fluctuations in superconducting circuits. en. Nat Commun 391

Appendix A
NRG—Implementation of the IRLM

Equations concerning the NRG implementation of the IRLM are given in this appendix, taking into account the conservation of the number of particles by the Hamiltonian to speed up calculations. The iterative construction of the Hamiltonian is firstly introduced, and in a second time the construction of some observables of interest is presented.

A.1 Flow of the Hamiltonian

As a reminder, two successive Hamiltonian in the Wilson chain construction are related as:

$$H_{N+1} = \sqrt{\Lambda} H_N + \sqrt{\Lambda}^N t_N \left(c_N^\dagger c_{N+1} + c_{N+1}^\dagger c_N \right), \tag{A.1}$$

where N labels here a given step of the flow, while n will denote a sector of n particle. It is admitted that H_N is already diagonal, with eigenstates written as $|\alpha, n\rangle_N$ for state α in the sector of n particles.

It is important for numerical simulations to know the size of each sector, to allocate the correct amount of memory. At step N, there are at most $n = N$ particles in the system. For each sector n, there are $A_{N-1,n}$ states α in the basis of last RG step. In this scheme, the truncation procedure works by sector: $A_{N,n}$ is at most N_{kept}, and every state with an energy above the largest kept energy, $\min\left(E_{N,n}^{N_{\text{kept}}}\right)$ in each sector n, is thrown away. The total number of kept states is thus: $N_{\text{kept,tot}} = \sum_n^N A_{N,n}$.

When H_{N+1} is built, a fermion $c_{N+1}^\dagger$ is added to the system. The new basis can be written:

M. Debertolis, *Quantum Impurity Problems in the Framework of Natural Orbitals*, Springer Theses, https://doi.org/10.1007/978-3-031-47233-6

$$\{(c_{N+1})^{\sigma}|\alpha, n\rangle_N\}, \quad \begin{cases} \sigma = 0, & \text{if } c_{N+1}^{\dagger} \text{ is not in the new state} \\ \sigma = 1, & \text{otherwise.} \end{cases} \tag{A.2}$$

To write the new basis for H_{N+1} in each sector of n particles, the introduction the new states that form a complete basis, in which the Hamiltonian is not diagonal, is done via the identity:

$$\mathbb{1} = \sum_{\sigma}^{\min(1,n)} \sum_{\beta}^{A_{N,n-\sigma}} (c_{N+1}^{\dagger})^{\sigma}|\beta, n-\sigma\rangle_N \langle\beta, n-\sigma|(c_{N+1})^{\sigma}, \tag{A.3}$$

which is used to write the unknown new eigenstates as:

$$|\alpha, n\rangle_{N+1} = \sum_{\sigma}^{\min(1,n)} \sum_{\beta}^{A_{N,n-\sigma}} (c_{N+1}^{\dagger})^{\sigma}|\beta, n-\sigma\rangle_N \langle\beta, n-\sigma|(c_{N+1})^{\sigma}|\alpha, n\rangle_{N+1}. \tag{A.4}$$

In the first sum, $\min(1, n)$ is used for the extreme cases $n = 0$ and $n = N + 1$: if $n = 0$, σ must be zero, since the sector $n = -1$ in the old basis makes no sense. For $n = N + 1$, σ must be 1, since there is no sector $n = N + 1$ in the basis of step N. With this form for the unknown $|\alpha, n\rangle_{N+1}$, the following matrix elements have been introduced:

$$_N\langle\beta, n-\sigma|(c_{N+1})^{\sigma}|\alpha, n\rangle_{N+1}, \tag{A.5}$$

and the left part is already known from the last iteration, but not the right one. We write the Hamiltonian in the new basis (in which H_{N+1} is diagonal, by definition), using Eq. (A.2):

$$\begin{aligned} \left[H_{N+1,n}\right]_{\alpha\sigma,\beta\sigma'} &= {}_N\langle\alpha, n-\sigma|(c_{N+1})^{\sigma} H_{N+1,n} (c_{N+1}^{\dagger})^{\sigma'}|\beta, n-\sigma'\rangle_N \\ &= {}_N\langle\alpha, n-\sigma|(c_{N+1})^{\sigma}\sqrt{\Lambda} H_N (c_{N+1}^{\dagger})^{\sigma'}|\beta, n-\sigma'\rangle_N \\ &+ \sqrt{\Lambda}^{N} t_N {}_N\langle\alpha, n-\sigma|(c_{N+1})^{\sigma}\left(c_N^{\dagger}c_{N+1} + c_{N+1}^{\dagger}c_N\right)(c_{N+1}^{\dagger})^{\sigma'}|\beta, n-\sigma'\rangle_N. \end{aligned} \tag{A.6}$$

The first term, that is diagonal in the sectors σ, σ' and in the states α, β, since it is the diagonal basis of H_N, reads:

$$_N\langle\alpha, n-\sigma|(c_{N+1})^{\sigma}\sqrt{\Lambda} H_N (c_{N+1}^{\dagger})^{\sigma'}|\beta, n-\sigma'\rangle_N = \delta_{\alpha\beta}\delta_{\sigma\sigma'}\sqrt{\Lambda} E_{N,n}^{\alpha}, \tag{A.7}$$

and the off-diagonal term:

$$\sqrt{\Lambda}^{N} t_{NN}\langle\alpha, n-\sigma|(c_{N+1})^{\sigma}\left(c_N^{\dagger}c_{N+1}+c_{N+1}^{\dagger}c_N\right)(c_{N+1}^{\dagger})^{\sigma'}|\beta, n-\sigma'\rangle_N$$
$$= \sqrt{\Lambda}^{-N} t_N \begin{cases} {}_N\langle\alpha, n|\, c_N^{\dagger}\, |\beta, n-1\rangle_N, & \text{if } \sigma=0, \sigma'=1 \\ {}_N\langle\alpha, n-1|\, c_N\, |\beta, n\rangle_N, & \text{if } \sigma=1, \sigma'=0 \\ 0 \ \text{otherwise.} \end{cases} \tag{A.8}$$

These matrix elements, unlike the ones in Eq. (A.5), are expressed with eigenvectors of the last step, that are known. However, what is known is not exactly the set of eigenvectors of last step, but the transformation matrix between the step $N-1$ and N. These elements are introduced through the identity $\mathbb{1} = \sum_{\mu} |\mu, n\rangle_{N-1}\langle\mu, n| + c_N^{\dagger}|\mu, n-1\rangle_{N-1}\langle\mu, n-1|c_N$:

$$\begin{aligned} {}_N\langle\alpha, n-1|c_N|\beta, n\rangle_N &= \sum_{\mu}^{A_{N-1,n-1}} {}_N\langle\alpha, n-1|\mu, n-1\rangle_{N-1}\langle\mu, n-1|c_N|\beta, n\rangle_N \\ &\quad + {}_N\langle\alpha, n-1|c_N^{\dagger}|\mu, n-2\rangle_{N-1}\langle\mu, n-2|c_N c_N|\beta, n\rangle_N \\ &= {}_N\langle\alpha, n-1|\mu, n-1\rangle_{N-1}\langle\mu, n-1|c_N|\beta, n\rangle_N. \end{aligned} \tag{A.9}$$

This form proposes to use two new matrices defined as:

$$\begin{aligned} \left[X_{N,n}\right]_{\alpha\beta} &= {}_{N-1}\langle\alpha, n|\beta, n\rangle_N \\ \left[Y_{N,n}\right]_{\alpha\beta} &= {}_{N-1}\langle\alpha, n-1|c_N|\beta, n\rangle_N. \end{aligned} \tag{A.10}$$

The dimension of $X_{N,n}$ is $\left(A_{N-1,n}, A_{N,n}\right)$, and $Y_{N,n}$ is of dimension $\left(A_{N-1,n-1}, A_{N,n}\right)$, and these two matrices allow to compute $H_{N+1,n}$ efficiently. It is clearer to show the matrix form of the Hamiltonian directly:

$$H_{N+1,n} = \begin{array}{c} \\ \sigma'=0 \\ \\ \sigma'=1 \end{array} \left[\begin{array}{c|c} \sigma=0 & \sigma=1 \\ \sqrt{\Lambda}\,\mathrm{diag}\left(E_{N,n}\right) & \sqrt{\Lambda}^{-N} t_N\, Y_{N,n}^{\dagger}\cdot X_{N,n-1} \\ \hline \sqrt{\Lambda}^{-N} t_N\, X_{N,n-1}^{\dagger}\cdot Y_{N,n} & \sqrt{\Lambda}\,\mathrm{diag}\left(E_{N,n-1}\right) \end{array} \right]. \tag{A.11}$$

Then, the Hamiltonian is diagonalized sector by sector, and we take the transformation matrix $P_{N,n}$ to define the new matrices $X_{N+1,n}$ and $Y_{N+1,n}$:

$$P_{N+1,n} = \begin{pmatrix} X_{N+1,n} \\ \\ Y_{N+1,n} \end{pmatrix} \begin{matrix} \} A_{N,n} \text{ states} \\ , \\ \} A_{N,n-1} \text{ states} \end{matrix} \tag{A.12}$$

and this transformation is repeated until the end of the flow. As it is a recursive process, there must be a first step. The initialization is done through H_0 containing the impurity degrees or freedom (and the interaction), which is diagonalized in each sector to start the flow. Once the flow of eigenvalues is running properly, observables can be computed in parallel.

A.2 Computing an Observable

The computation of any static observable concerning the impurity or the fermions of the discretized chain of the IRLM is presented here. Any observable of this kind is defined through an operator $\mathcal{O}$, and it is considered that it changes the number of particle of a state by m. If the operator is acts on fermions that are already in the flow (if it concerns the impurity, it is defined on the first step), the observable has to be defined for all pair of states (α, β) to follow the flow of eigenvectors, and it is evaluated in the ground state at the end. At a given iteration, it reads:

$$\begin{aligned} \left[\mathcal{O}_{N+1,n}\right]_{\alpha\beta} &= {}_{N+1}\langle \alpha, n-m | \mathcal{O}_{N+1,n} | \beta, n \rangle_{N+1} \\ &= \begin{pmatrix} X^{\dagger}_{N+1,n-m} & Y^{\dagger}_{N+1,n-m} \end{pmatrix} \begin{pmatrix} \mathcal{O}_{11} & \mathcal{O}_{12} \\ \mathcal{O}_{21} & \mathcal{O}_{22} \end{pmatrix} \begin{pmatrix} X_{N+1,n} \\ Y_{N+1,n} \end{pmatrix}. \end{aligned} \tag{A.13}$$

The usefulness of using matrices X and Y is evident here, the expressions are simple to write. It is again supposed here that the observable is known in the last basis $\mathcal{O}_{N,n}$. Hence, each sector of $\mathcal{O}_{N+1,n}$ reads:

$$\begin{aligned} \left[\mathcal{O}_{11}\right]_{\alpha\beta} &= {}_{N}\langle \alpha, n-m | \mathcal{O}_{N+1,n} | \beta, n \rangle_{N} \\ \left[\mathcal{O}_{12}\right]_{\alpha\beta} &= {}_{N}\langle \alpha, n-m | \mathcal{O}_{N+1,n-1}\, c^{\dagger}_{N+1} | \beta, n-1 \rangle_{N} \\ \left[\mathcal{O}_{21}\right]_{\alpha\beta} &= {}_{N}\langle \alpha, n-1-m | c_{N+1}\, \mathcal{O}_{N+1,n} | \beta, n \rangle_{N} \\ \left[\mathcal{O}_{22}\right]_{\alpha\beta} &= {}_{N}\langle \alpha, n-1-m | c_{N+1}\, \mathcal{O}_{N+1,n-1}\, c^{\dagger}_{N+1} | \beta, n-1 \rangle_{N}. \end{aligned} \tag{A.14}$$

For the general case, we can not go further with the equations, and we will give here the details of the observable required for the study of natural orbitals in the IRLM, the one-body density matrix Q.

A.3 One-Body Density Matrix

We recall that the density matrix is defined as one-body equal-time correlators for every pair of fermions in the system:

$$\left[Q_{ij,N,n}\right]_{\alpha\beta} = {}_N\langle\alpha, n|c_i^\dagger c_j|\beta, n\rangle_N. \tag{A.15}$$

It has four entries along the flow in each sector of fixed number of particle n, since each element of Q is an observable as defined above. At the end of the flow, the one-body density matrix is diagonalized with each of its elements evaluated in the ground state. Each element of Q is an observable defined from operators conserving the number of particles, such that $m = 0$ here. At each step of the flow, the fermion $c_N^\dagger$ is added to the system, such that the square matrix Q is always growing and of dimension N. Elements bearing c_N should be taken apart, since this fermion appears explicitely in $|\beta, n\rangle_{N+1}$ and thus they do not (anti-)commute. We will thus separate the evolution of elements $i, j < N$, $i = N, j = N$ and $i = j = N$.

a. Case $i, j < N + 1$
We solve Eq. (A.14) for the elements of matrix Q that are not in the border. It is plain that as $c_{N+1}^\dagger$ is not in the operators that we consider here, $Q_{12} = Q_{21} = 0$, since $c_{N+1}^\dagger$ appears only once and the projection on the two resulting states is trivially zero. For Q_{11} and Q_{22}, there is an equal number of creation and annihilation operators involving the $N + 1$ fermions, such that they are non-zero and read:

$$\begin{aligned} Q_{ij,N+1,n,11,\alpha\beta} &= {}_N\langle\alpha, n|c_i^\dagger c_j|\beta, n\rangle_N = Q_{ij,N,n,\alpha\beta} \\ Q_{ij,N+1,n,22,\alpha\beta} &= {}_N\langle\alpha, n-1|c_i^\dagger c_j|\beta, n-1\rangle_N = Q_{ij,N,n-1,\alpha\beta}, \end{aligned} \tag{A.16}$$

such that the new operator is defined from the operator at the previous step as:

$$Q_{ij,N+1,n} = X_{N+1,n}^\dagger Q_{ij,N,n} X_{N+1,n} + Y_{N,n}^\dagger Q_{ij,N,n-1} Y_{N,n}. \tag{A.17}$$

Hence, $Q_{N+1,n}$ is easily updated along the flow after computing the Hamiltonian eigenvectors, and by storing the $Q_{N,n}$ of the previous iteration. In the following, the same arguments are used and we just give the results for $Q_{N+1,n}$.

b. Case $i = N + 1, j < N + 1$

$$Q_{N+1j,N+1,n} = Y_{N+1,n}^\dagger {}_N\langle\alpha, n-1|\, c_j \,|\beta, n\rangle_N X_{N+1,n}. \tag{A.18}$$

In this case, it appears that only the sector Q_{21} is not zero, and the $c^\dagger_{N+1}$ of Q cancel with the one of the state on which we project it. Hence, only a c_j survives, that is not calculated yet. It means that we have to follow the vector constituted of all operators c_j along the flow to compute Q. It will be labeled $c_{j,N,n}$, and it is computed just after the last cases for Q. With this notation, the last equation becomes:

$$Q_{N+1j,N+1,n} = Y^\dagger_{N+1,nN} c_{j,N,n} X_{N+1,n}. \tag{A.19}$$

c. Case $i < N+1,\ j = N+1$

$$Q_{iN+1,N+1,n} = X^\dagger_{N+1,nN} c^\dagger_{i,N,n} Y_{N+1,n}. \tag{A.20}$$

$c^\dagger_{N,n,i}$ is not computed separately and is defined as the hermitian conjugate of $c_{i,N,n}$.

d. Case $i = j = N+1$

$$Q_{N+1N+1,N+1,n} = Y^\dagger_{N+1,n} Y_{N+1,n}. \tag{A.21}$$

A.3.1 Annihilation Operators

As for Q, the case c_{N+1} is treated separately. This operator changes the number of fermions, and we take here $m = 1$.

a. Case $i < N+1$

$$c_{i,N+1,n} = X^\dagger_{N+1,n-1} c_{i,N,n} X_{N+1,n} - Y^\dagger_{N+1,n-1} c_{i,N,n-1} Y_{N+1,n}. \tag{A.22}$$

b. Case $i = N+1$

$$c_{N+1,N+1,n} = X^\dagger_{N+1,n-1} Y_{N+1,n}. \tag{A.23}$$

Every equation is now closed, and the only thing left to do is the initialization of Q and c at the first step of the NRG procedure, with H_0. Then, the truncation of high energy states is only a change of each number of states per sector $A_{N,n}$, and the thrown states only change the dimension of matrices $X_{N,n}$ and $Y_{N,n}$, that will affect the observable automatically. For the truncation, we find the sector in which the N_{kept} state has the lowest energy (if there are at least this amount of states in the given sector), that is usually the half-filled sector at particle-hole symmetry. Then, every state with an higher energy in every other sector is thrown away. Another small thing not to forgot is to shift the spectrum by the lowest energy $E^0_{N,n}$, for the rescaling by $\sqrt{\Lambda}$ to be properly done.

A.4 Initialization

A.4.1 Hamiltonian

The Hamiltonian considered before iterating is composed of the impurity and of the first site of the chain:

$$H_0 = U\left(d^\dagger d - \frac{1}{2}\right)\left(c_0^\dagger - \frac{1}{2}\right) + V\left(d^\dagger c_0 + \text{h.c.}\right) + \epsilon_d d^\dagger d \tag{A.24}$$

We will directly work by sector of states with equal number of particles, to define the transfer matrices as defined above. There can be 0, 1 or 2 particle in this small system, hence $N = 2$ corresponds to the initialization step, so that t_N must be taken carefully: when $N = 3$ corresponding to the maximum number of particles, the hopping term is t_0 and not t_3. By sector, the states are:

∘ $n = 0$:

$$\text{One state: } |0\rangle, \quad E_{20} = \Lambda^{-0.5}\frac{U}{4} \rightarrow X_{20} = ((1)), Y_{20} = (()). \tag{A.25}$$

∘ $n = 1$:

$$H_{21} = \Lambda^{-0.5}\begin{pmatrix} d^\dagger\,|0\rangle & c^\dagger\,|0\rangle \\ \varepsilon_d - \frac{U}{4} & V \\ V & -\frac{U}{4} \end{pmatrix} \rightarrow \text{Diagonalization}: P = \begin{pmatrix} X_{21} \\ X_{21} \end{pmatrix}. \tag{A.26}$$

∘ $n = 2$:

$$\text{One state: } d^\dagger c_0^\dagger|0\rangle, \quad E_{22} = \Lambda^{-0.5}\left(\frac{U}{4} + \epsilon_d\right) \rightarrow X_{22} = (()), Y_{22} = ((1)). \tag{A.27}$$

A.4.2 Q Matrix and c Vector

We use the initialization of X_{2n} and Y_{2n} for operators of $Q_{ij,2n}$ and the annihilation operators $c_{i,2n}$:

∘ $n = 0$:

$$Q_{20} = \begin{pmatrix} 0 & 0 \\ 0 & 0 \end{pmatrix}, \quad c_{20} = \begin{pmatrix} 0 \\ 0 \end{pmatrix}. \tag{A.28}$$

$on = 1$:

$$Q_{21} = \begin{pmatrix} X_{21}^{\dagger} X_{21} & Y_{21}^{\dagger} X_{21} \\ X_{21}^{\dagger} Y_{21} & Y_{21}^{\dagger} Y_{21} \end{pmatrix}, \quad c_{21} = \begin{pmatrix} X_{20}^{\dagger} X_{21} \\ X_{20}^{\dagger} Y_{21} \end{pmatrix}. \tag{A.29}$$

$on = 2$:

$$Q_{22} = \begin{pmatrix} 1 & 0 \\ 0 & 1 \end{pmatrix}, \quad c_{22} = \begin{pmatrix} -Y_{21}^{\dagger} Y_{22} \\ X_{21}^{\dagger} Y_{22} \end{pmatrix}. \tag{A.30}$$

It is recalled that any entry of Q and c correponds to the matrix form of operator in a given sector. An integer in Q of c makes no sense, it has to be understood as an operator defined in a basis composed of one state, and is written only to simplify notations. Any static observable concerning only the impurity or the first site can be defined at this moment, and then follows the flow of eigenvectors as explained above.

We provided all details of the NRG implementation of the IRLM, and of its one-body density matrix (more generally of any static observable). In Appendix B, we will provide the same equations for the SIAM, with an additional symmetry involving the conservation of the spin.

Appendix B
NRG—Implementation of the SIAM

As in Appendix A for the IRLM, the NRG implementation of the SIAM is discussed in this appendix. The implementation accounts for the charge and spin conservations, that is more involved than only charge conservation in the IRLM. Also, we will give less details when the calculations are similar to those of Appendix A, so we recommand that the reader read it before this one. As in Appendix A, the flow of eigenvectors of the SIAM will be firstly derived, and followed by the calculation of the one-body density matrix of this model.

B.1 Flow of the Hamiltonian

The iterative construction of the SIAM, which considers spinful fermions, reads:

$$H_{N+1} = \sqrt{\Lambda} H_N + \sqrt{\Lambda}^N t_N \left(c^\dagger_{\uparrow N} c_{\uparrow N+1} + c^\dagger_{\downarrow N} c_{\downarrow N} + \text{h.c.} \right), \tag{B.1}$$

such that two fermions are added at each step, to respect the conservation of the spin. The new basis for H_{N+1}, knowing the one that diagonalizes H_N, reads:

$$|\alpha, n, s_z\rangle_{N+1} = \sum_{ab}^{1} (c^\dagger_{\uparrow N+1})^a \otimes (c^\dagger_{\uparrow N+1})^b \otimes |\alpha, n-(a+b), s_z-(a-b)\rangle_N, \tag{B.2}$$

where $n = n_\uparrow + n_\downarrow$ ($a + b$ for the new fermions) is the charge, and $n = n_\uparrow - n_\downarrow$ ($a - b$) the spin. The Hamiltonian is defined by block for each fixed particle number n and spin s_z. As before, the number of states in each block reads A_{N,n,s_z}. As before, the Hamiltonian is projected in the new basis:

M. Debertolis, *Quantum Impurity Problems in the Framework of Natural Orbitals*, Springer Theses, https://doi.org/10.1007/978-3-031-47233-6

$$\left[H_{N+1,n,s_z}\right]_{\alpha\beta} = \sum_{abcd}^{1} {}_N\langle \alpha, n-(a+b), s_z-(a-b)|\,(c_{\downarrow N+1})^b (c_{\uparrow N+1})^a \times$$
$$\times\, H_{N+1,n,s_z} (c^\dagger_{\uparrow N+1})^c (c^\dagger_{\downarrow N+1})^d\, |\beta, n-(c+d), s_z-(c-d)\rangle_N. \tag{B.3}$$

It is clear from Eq. (B.1) that some combinations of a, b, c and d can not exist, if numbers of c_{N+1} and $c^\dagger_{N+1}$ in Eq. (B.3) are not the same. As before, the Hamitonian is written in a matrix form for readability:

$$H_{N+1,n,s_z} = \begin{array}{cc} & \begin{array}{cccc} \multicolumn{2}{c}{c=1} & \multicolumn{2}{c}{c=0} \\ d=1 & d=0 & d=1 & d=0 \end{array} \\ \begin{array}{cc} a=1 & \begin{array}{c} b=1 \\ b=0 \end{array} \\ a=0 & \begin{array}{c} b=1 \\ b=0 \end{array} \end{array} & \left(\begin{array}{cc|cc} (a) & (1) & (2) & 0 \\ & (b) & 0 & (3) \\ \hline & 0 & (c) & (4) \\ 0 & & & (d) \end{array}\right) \end{array} \tag{B.4}$$

Diagonal terms (a), (b), (c) and (d) are constituted of the eigenvalues of H_N, while the off-diagonal ones are coming from the hopping terms of Eq. (B.1), and blank cases correspond to hermitian conjugates of (1), (2), (3) and (4). The corresponding matrix elements reads:

$$\begin{aligned}
[(a)]_{\alpha,\beta} &= \delta_{\alpha\beta} \sqrt{\Lambda}\, E^\alpha_{N,n-2,s_z} \\
[(b)]_{\alpha,\beta} &= \delta_{\alpha\beta} \sqrt{\Lambda}\, E^\alpha_{N,n-1,s_z-1} \\
[(c)]_{\alpha,\beta} &= \delta_{\alpha\beta} \sqrt{\Lambda}\, E^\alpha_{N,n-1,s_z+1} \\
[(d)]_{\alpha,\beta} &= \delta_{\alpha\beta} \sqrt{\Lambda}\, E^\alpha_{N,n,s_z} \\
[(1)]_{\alpha,\beta} &= \sqrt{\Lambda}^N\, t_N\, {}_N\langle \alpha, n-2, s_z|\, c_{\downarrow N}\, |\beta, n-1, s_z-1\rangle_N \\
[(2)]_{\alpha,\beta} &= -\sqrt{\Lambda}^N\, t_N\, {}_N\langle \alpha, n-2, s_z|\, c_{\uparrow N}\, |\beta, n-1, s_z+1\rangle_N \\
[(3)]_{\alpha,\beta} &= \sqrt{\Lambda}^N\, t_N\, {}_N\langle \alpha, n-1, s_z-1|\, c_{\uparrow N}\, |\beta, n, s_z\rangle_N \\
[(4)]_{\alpha,\beta} &= \sqrt{\Lambda}^N\, t_N\, {}_N\langle \alpha, n-1, s_z+1|\, c_{\downarrow N}\, |\beta, n, s_z\rangle_N.
\end{aligned} \tag{B.5}$$

At this point, the transformation matrix between step $N-1$ and N is introduced for the off-diagonal terms, as before, via the completeness relations (the energies are omitted here for readability, they will appear back later):

$$
\begin{aligned}
[(1)]_{\alpha,\beta} = \sum_{\gamma} {}_N\langle \alpha, n-2, s_z | \gamma, n-2, s_z \rangle_{N-1} \langle \gamma, n-2, s_z | c_{\downarrow N} | \beta, n-1, s_z-1 \rangle_N \\
+ {}_N\langle \alpha, n-2, s_z | c^\dagger_{\uparrow N} | \gamma, n-3, s_z-1 \rangle_{N-1} \langle \gamma, n-3, s_z-1 | c_{\uparrow N} c_{\downarrow N} | \beta, n-1, s_z-1 \rangle_N \\
+ {}_N\langle \alpha, n-2, s_z | c^\dagger_{\downarrow N} | \gamma, n-3, s_z+1 \rangle_{N-1} \langle \gamma, n-3, s_z+1 | c_{\downarrow N} c_{\downarrow N} | \beta, n-1, s_z-1 \rangle_N \\
+ {}_N\langle \alpha, n-2, s_z | c^\dagger_{\uparrow N} c^\dagger_{\downarrow N} | \gamma, n-4, s_z-1 \rangle_{N-1} \langle \gamma, n-4, s_z | c_{\downarrow N} c_{\uparrow N} c_{\downarrow N} | \beta, n-1, s_z-1 \rangle_N \\
[(2)]_{\alpha,\beta} = \sum_{\gamma} {}_N\langle \alpha, n-2, s_z | \gamma, n-2, s_z \rangle_{N-1} \langle \gamma, n-2, s_z | c_{\uparrow N} | \beta, n-1, s_z+1 \rangle_N + \dots \\
[(3)]_{\alpha,\beta} = \sum_{\gamma} {}_N\langle \alpha, n-1, s_z-1 | \gamma, n-1, s_z-1 \rangle_{N-1} \langle \gamma, n-1, s_z-1 | c_{\uparrow N} | \beta, n, s_z \rangle_N + \dots \\
[(4)]_{\alpha,\beta} = \sum_{\gamma} {}_N\langle \alpha, n-1, s_z+1 | \gamma, n-1, s_z+1 \rangle_{N-1} \langle \gamma, n-1, s_z+1 | c_{\downarrow N} | \beta, n, s_z \rangle_N + \dots,
\end{aligned}
\tag{B.6}
$$

and cancelling each zero term, it simplifies to:

$$
\begin{aligned}
[(1)]_{\alpha,\beta} = \sum_{\gamma} {}_N\langle \alpha, n-2, s_z | \gamma, n-2, s_z \rangle_{N-1} \langle \gamma, n-2, s_z | c_{\downarrow N} | \beta, n-1, s_z-1 \rangle_N \\
+ {}_N\langle \alpha, n-2, s_z | c^\dagger_{\uparrow N} | \gamma, n-3, s_z-1 \rangle_{N-1} \langle \gamma, n-3, s_z-1 | c_{\uparrow N} c_{\downarrow N} | \beta, n-1, s_z-1 \rangle_N \\
[(2)]_{\alpha,\beta} = \sum_{\gamma} {}_N\langle \alpha, n-2, s_z | \gamma, n-2, s_z \rangle_{N-1} \langle \gamma, n-2, s_z | c_{\uparrow N} | \beta, n-1, s_z+1 \rangle_N \\
+ {}_N\langle \alpha, n-2, s_z | c^\dagger_{\downarrow N} | \gamma, n-3, s_z+1 \rangle_{N-1} \langle \gamma, n-3, s_z+1 | c_{\downarrow N} c_{\uparrow N} | \beta, n-1, s_z+1 \rangle_N \\
[(3)]_{\alpha,\beta} = \sum_{\gamma} {}_N\langle \alpha, n-1, s_z-1 | \gamma, n-1, s_z-1 \rangle_{N-1} \langle \gamma, n-1, s_z-1 | c_{\uparrow N} | \beta, n, s_z \rangle_N \\
+ {}_N\langle \alpha, n-1, s_z-1 | c^\dagger_{\downarrow N} | \gamma, n-2, s_z \rangle_{N-1} \langle \gamma, n-2, s_z | c_{\downarrow N} c_{\uparrow N} | \beta, n, s_z \rangle_N \\
[(4)]_{\alpha,\beta} = \sum_{\gamma} {}_N\langle \alpha, n-1, s_z+1 | \gamma, n-1, s_z+1 \rangle_{N-1} \langle \gamma, n-1, s_z+1 | c_{\downarrow N} | \beta, n, s_z \rangle_N \\
+ {}_N\langle \alpha, n-1, s_z+1 | c^\dagger_{\uparrow N} | \gamma, n-2, s_z \rangle_{N-1} \langle \gamma, n-2, s_z | c_{\uparrow N} c_{\downarrow N} | \beta, n, s_z \rangle_N,
\end{aligned}
\tag{B.7}
$$

It is natural here to introduce blocks of the transformation matrix, that reads:

$$
P_{N,n,s_z} = \begin{pmatrix} D_{N,n,s_z} \\ C_{N,n,s_z} \\ B_{N,n,s_z} \\ A_{N,n,s_z} \end{pmatrix}
\begin{matrix} \} c_{\uparrow N} c_{\downarrow N} \\ \} c_{\uparrow N} \\ \} c_{\downarrow N} \\ \} \end{matrix} .
\tag{B.8}
$$

Matrices A, B, C and D are written explicitely as:

$$\begin{aligned}
\left[A_{N,n,s_z}\right]_{\alpha\beta} &= {}_{N-1}\langle\alpha, n, s_z|\beta, n, s_z\rangle_N \\
\left[B_{N,n,s_z}\right]_{\alpha\beta} &= {}_{N-1}\langle\alpha, n-1, s_z+1|\, c_{\downarrow N}\, |\beta, n, s_z\rangle_N \\
\left[C_{N,n,s_z}\right]_{\alpha\beta} &= {}_{N-1}\langle\alpha, n-1, s_z-1|\, c_{\uparrow N}\, |\beta, n, s_z\rangle_N \\
\left[D_{N,n,s_z}\right]_{\alpha\beta} &= {}_{N-1}\langle\alpha, n-2, s_z|\, c_{\uparrow N}c_{\downarrow N}\, |\beta, n, s_z\rangle_N,
\end{aligned} \tag{B.9}$$

which allow to write the set of Eq. (B.7) in a simpler form:

$$\begin{aligned}
[(1)]_{\alpha,\beta} &= \sqrt{\Lambda}^N t_N \left(A^\dagger_{N,n-2,s_z}\cdot B_{N,n-1,s_z-1} - C^\dagger_{N,n-2,s_z}\cdot D_{N,n-1,s_z-1}\right) \\
[(2)]_{\alpha,\beta} &= -\sqrt{\Lambda}^N t_N \left(A^\dagger_{N,n-2,s_z}\cdot C_{N,n-1,s_z-1} - B^\dagger_{N,n-2,s_z}\cdot D_{N,n-1,s_z+1}\right) \\
[(3)]_{\alpha,\beta} &= \sqrt{\Lambda}^N t_N \left(A^\dagger_{N,n-1,s_z-1}\cdot C_{N,n,s_z} - B^\dagger_{N,n-1,s_z-1}\cdot D_{N,n,s_z}\right) \\
[(4)]_{\alpha,\beta} &= \sqrt{\Lambda}^N t_N \left(A^\dagger_{N,n-1,s_z+1}\cdot B_{N,n,s_z} - C^\dagger_{N,n-1,s_z+1}\cdot D_{N,n,s_z}\right).
\end{aligned} \tag{B.10}$$

Now that Hamiltonian (B.4) is fully determined, it is diagonalized by blocks, into which new matrices A, B, C and D are identified via P_{N+1,n,s_z}. The truncation procedure is the same as explained in Appendix A. It was not detailed here, but as in the IRLM, the number of particles in each sector A_{N,n,s_z} has to be followed for an efficient use of the computer memory. Now that the flow of eigenvectors is defined, we present here how to compute observables, and, as the calculations are more involved than in the previous appendix, we will focus directly on the one-body density matrix and skip the general case (note that it is not more difficult to calculate, but this lenghty derivation is not of great interest here).

B.2 One-Body Density Matrix

In the spinless case, spins up and down are stacked in different blocks of the matrix. In the diagonal blocks, operators are $Q_{ij,\sigma\sigma} = \langle c^\dagger_{\sigma i}c_{\sigma j}\rangle$, and they conserve both the number of particles n and the spin s_z. That is not the case for off-diagonal blocks, that are in $\sigma\sigma'$ and do not conserve s_z anymore. As for the IRLM, operators c_{i,σ,n,s_z} are required for the flow of $Q_{i,j,\sigma,\sigma',n,s_z}$, and they neither conserve n nor s_z. The flow of annihilation operators is discussed here also.

Definition of an Operator

In the eigenbasis of step $N+1$, an operator $\mathcal{O}$ changing the n by m and s_z by s reads:

$$\left[\mathcal{O}_{N+1,n}\right]_{\alpha\beta} = {}_{N+1}\langle\alpha, n-m|\mathcal{O}_{N+1,n}|\beta, n\rangle_{N+1}$$

$$= \left(D^\dagger\ C^\dagger\ B^\dagger\ A^\dagger\right)_{N+1,n-m,s_z-s} \begin{pmatrix} \mathcal{O}_{11} & \mathcal{O}_{12} & \mathcal{O}_{13} & \mathcal{O}_{14} \\ \mathcal{O}_{21} & \mathcal{O}_{22} & \mathcal{O}_{23} & \mathcal{O}_{24} \\ \mathcal{O}_{31} & \mathcal{O}_{32} & \mathcal{O}_{33} & \mathcal{O}_{34} \\ \mathcal{O}_{41} & \mathcal{O}_{42} & \mathcal{O}_{43} & \mathcal{O}_{44} \end{pmatrix} \begin{pmatrix} D_{N+1,n,s_z} \\ C_{N+1,n,s_z} \\ B_{N+1,n,s_z} \\ A_{N+1,n,s_z} \end{pmatrix}. \tag{B.11}$$

Operators that we will consider are hermitian, so we only consider blocks above the diagonal. As an example, the element $\mathcal{O}_{12}$ reads:

$$\left[\mathcal{O}_{12}\right]_{\alpha\beta} = {}_N\langle\alpha, n-m-2, s_z - s|\, c_{\downarrow N+1} c_{\uparrow N+1} \mathcal{O}_N c^\dagger_{\uparrow N+1}\, |\beta, n-1, s_z - 1\rangle_N. \tag{B.12}$$

The actual calculations for the operators of the reduced density matrix are given now.

Operators $\mathcal{O} = c^\dagger_{\sigma i} c_{\sigma j}$: $m = s = 0$

- $i, j < N+1$: terms $\neq 0 \rightarrow \mathcal{O}_{11}, \mathcal{O}_{22}, \mathcal{O}_{33}, \mathcal{O}_{44}$

$$\begin{aligned} \mathcal{O}_{i,j,N+1,n,s_z} &= D^\dagger_{N+1,n,s_z} \mathcal{O}_{i,j,N,n-2,s_z} D_{N+1,n,s_z} + C^\dagger_{N+1,n,s_z} \mathcal{O}_{i,j,N,n-1,s_z-1} C_{N+1,n,s_z} \\ &+ B^\dagger_{N+1,n,s_z} \mathcal{O}_{i,j,N,n-1,s_z+1} B_{N+1,n,s_z} + A^\dagger_{N+1,n,s_z} \mathcal{O}_{i,j,N,n,s_z} A_{N+1,n,s_z} \end{aligned} \tag{B.13}$$

- $i = N+1$, $j < N+1$ and $\sigma = \uparrow$: terms $\neq 0 \rightarrow \mathcal{O}_{13}, \mathcal{O}_{24}$

$$\mathcal{O}_{i,j,N+1,n,s_z} = -D^\dagger_{N+1,n,s_z} c_{\uparrow j,N,n-1,s_z+1} B_{N+1,n,s_z} + C^\dagger_{N+1,n,s_z} c_{\uparrow j,N,n,s_z} A_{N+1,n,s_z} \tag{B.14}$$

- $i = N+1$, $j < N+1$ and $\sigma = \downarrow$: terms $\neq 0 \rightarrow \mathcal{O}_{12}, \mathcal{O}_{34}$

$$\mathcal{O}_{i,j,N+1,n,s_z} = D^\dagger_{N+1,n,s_z} c_{\downarrow j,N,n-1,s_z-1} C_{N+1,n,s_z} + B^\dagger_{N+1,n,s_z} c_{\downarrow j,N,n,s_z} A_{N+1,n,s_z} \tag{B.15}$$

- $i = j = N+1$ and $\sigma = \uparrow$: terms $\neq 0 \rightarrow \mathcal{O}_{11}, \mathcal{O}_{22}$

$$\mathcal{O}_{i,j,N+1,n,s_z} = D^\dagger_{N+1,n,s_z} D_{N+1,n,s_z} + C^\dagger_{N+1,n,s_z} C_{N+1,n,s_z} \tag{B.16}$$

- $i = j = N+1$ and $\sigma = \downarrow$: terms $\neq 0 \rightarrow \mathcal{O}_{11}, \mathcal{O}_{33}$

$$\mathcal{O}_{i,j,N+1,n,s_z} = D^\dagger_{N+1,n,s_z} D_{N+1,n,s_z} + B^\dagger_{N+1,n,s_z} B_{N+1,n,s_z} \tag{B.17}$$

Operators $\mathcal{O} = c^\dagger_{\uparrow i} c_{\downarrow j}$: $m = 0$, $s = -2$

- $i, j < N+1$: terms $\neq 0 \rightarrow \mathcal{O}_{11}, \mathcal{O}_{22}, \mathcal{O}_{33}, \mathcal{O}_{44}$

$$\begin{aligned} \mathcal{O}_{i,j,N+1,n,s_z} &= D^\dagger_{N+1,n,s_z+2} \mathcal{O}_{i,j,N,n-2,s_z} D_{N+1,n,s_z} + C^\dagger_{N+1,n,s_z+2} \mathcal{O}_{i,j,N,n-1,s_z-1} C_{N+1,n,s_z} \\ &+ B^\dagger_{N+1,n,s_z+2} \mathcal{O}_{i,j,N,n-1,s_z+1} B_{N+1,n,s_z} + A^\dagger_{N+1,n,s_z+2} \mathcal{O}_{i,j,N,n,s_z} A_{N+1,n,s_z} \end{aligned} \tag{B.18}$$

∘ $i = N+1,\ j < N+1$: terms $\neq 0 \rightarrow \mathcal{O}_{13}, \mathcal{O}_{24}$

$$\mathcal{O}_{i,j,N+1,n,s_z} = -D^{\dagger}_{N+1,n,s_z+2}\, c_{\downarrow j,N,n-1,s_z+1}\, B_{N+1,n,s_z} + C^{\dagger}_{N+1,n,s_z+2}\, c_{\downarrow j,N,n,s_z}\, A_{N+1,n,s_z} \tag{B.19}$$

∘ $i < N+1,\ j = N+1$: terms $\neq 0 \rightarrow \mathcal{O}_{21}, \mathcal{O}_{43}$

$$\mathcal{O}_{i,j,N+1,n,s_z} = A^{\dagger}_{N+1,n,s_z+2}\, c^{\dagger}_{\uparrow j,N,n,s_z+2}\, B_{N+1,n,s_z} + C^{\dagger}_{N+1,n,s_z+2}\, c^{\dagger}_{\downarrow j,N,n-1,s_z+1}\, D_{N+1,n,s_z} \tag{B.20}$$

∘ $i = j = N+1$: terms $\neq 0 \rightarrow \mathcal{O}_{23}$

$$\mathcal{O}_{i,j,N+1,n,s_z} = C^{\dagger}_{N+1,n,s_z+2}\, B_{N+1,n,s_z} \tag{B.21}$$

Operators $\mathcal{O} = c_{\uparrow i}$: $m = 1,\ s = 1$

∘ $i < N+1$: terms $\neq 0 \rightarrow \mathcal{O}_{11}, \mathcal{O}_{22}, \mathcal{O}_{33}, \mathcal{O}_{44}$

$$\begin{aligned}\mathcal{O}_{i,j,N+1,n,s_z} &= D^{\dagger}_{N+1,n-1,s_z-1}\, \mathcal{O}_{i,j,N,n-2,s_z}\, D_{N+1,n,s_z} - C^{\dagger}_{N+1,n-1,s_z-1}\, \mathcal{O}_{i,j,N,n-1,s_z-1}\, C_{N+1,n,s_z}\\ &\quad - B^{\dagger}_{N+1,n-1,s_z-1}\, \mathcal{O}_{i,j,N,n-1,s_z+1}\, B_{N+1,n,s_z} + A^{\dagger}_{N+1,n-1,s_z-1}\, \mathcal{O}_{i,j,N,n,s_z}\, A_{N+1,n,s_z}\end{aligned} \tag{B.22}$$

∘ $i = N+1$: terms $\neq 0 \rightarrow \mathcal{O}_{31}, \mathcal{O}_{42}$

$$\mathcal{O}_{i,j,N+1,n,s_z} = A^{\dagger}_{N+1,n-1,s_z-1}\, C_{N+1,n,s_z} + B^{\dagger}_{N+1,n-1,s_z-1}\, D_{N+1,n,s_z} \tag{B.23}$$

Operators $\mathcal{O} = c_{\downarrow i}$: $m = 1,\ s = -1$

∘ $i < N+1$: terms $\neq 0 \rightarrow \mathcal{O}_{11}, \mathcal{O}_{22}, \mathcal{O}_{33}, \mathcal{O}_{44}$

$$\begin{aligned}\mathcal{O}_{i,j,N+1,n,s_z} &= D^{\dagger}_{N+1,n-1,s_z+1}\, \mathcal{O}_{i,j,N,n-2,s_z}\, D_{N+1,n,s_z} - C^{\dagger}_{N+1,n-1,s_z+1}\, \mathcal{O}_{i,j,N,n-1,s_z-1}\, C_{N+1,n,s_z}\\ &\quad - B^{\dagger}_{N+1,n-1,s_z+1}\, \mathcal{O}_{i,j,N,n-1,s_z+1}\, B_{N+1,n,s_z} + A^{\dagger}_{N+1,n-1,s_z+1}\, \mathcal{O}_{i,j,N,n,s_z}\, A_{N+1,n,s_z}\end{aligned} \tag{B.24}$$

∘ $i = N+1$: terms $\neq 0 \rightarrow \mathcal{O}_{21}, \mathcal{O}_{43}$

$$\mathcal{O}_{i,j,N+1,n,s_z} = A^{\dagger}_{N+1,n-1,s_z+1}\, B_{N+1,n,s_z} - C^{\dagger}_{N+1,n-1,s_z+1}\, D_{N+1,n,s_z} \tag{B.25}$$

This closes the recursive definition of the one-body density matrix of the SIAM, and only the initialization has to be done by hand before starting the flow. The truncation acts on the matrices A, ..., D, such that operators are at each step defined in the truncated basis.

B.3 Initialization

In the SIAM, the interaction part is local on the impurity, such that the initialization can only take into account the 2 fermions on the impurity. However, the hopping term V has to be added to the first step of the flow only. One can chose to start with the impurity and the first site, so that the flow is defined the same way at each step. The price to pay is the larger basis for the initialization. We make the first choice for simplicity here, but the second one was chosen for the actual implementation of the algorithm (the results are the same).

B.3.1 Hamiltonian

As in Appendix A, the initialization step starts at $N = 2$, and s_z ranges from -1 to 1.

∘ $n = 0$, $s_z = 0$:

$$\text{One state: } |0\rangle, \quad E_{200} = \Lambda^{-0.5}\frac{U}{4} \quad \rightarrow A_{200} = ((1)),\, B_{200} = C_{200} = D_{200} = (()). \tag{B.26}$$

∘ $n = 1$:

∘ $s_z = -1$:

$$\text{One state: } d_\downarrow^\dagger|0\rangle, \quad E_{21-1} = -\Lambda^{-0.5}\frac{U}{2} \quad \rightarrow C_{21-1} = ((1)),\, A_{21-1} = B_{21-1} = D_{21-1} = (()). \tag{B.27}$$

∘ $s_z = 0$: No states

∘ $s_z = 1$:

$$\text{One state: } d_\uparrow^\dagger|0\rangle, \quad E_{211} = -\Lambda^{-0.5}\frac{U}{2} \quad \rightarrow B_{211} = ((1)),\, A_{211} = C_{211} = D_{211} = (()). \tag{B.28}$$

∘ $n = 2$: No states for $s_z \neq 2$.

∘ $s_z = 2$:

$$\text{One state: } d_\uparrow^\dagger d_\uparrow^\dagger|0\rangle, \quad E_{222} = \Lambda^{-0.5}\frac{U}{4} \quad \rightarrow D_{222} = ((1)),\, A_{222} = B_{222} = C_{222} = (()). \tag{B.29}$$

Transformation matrices being initialized, we can initialize the operators of interest. At the first step of the flow, the off-diagonal elements to be considered are not proportional to t_0, but to V. Then, for $N = 6$, t_0 is introduced as in the above calculation.

B.3.2 Q Matrix and c Vector

With a similar format to Appendix A, the first definition of these operators are given, sector by sector: ∘ $n = 0, s_z = 0$:

$$Q_{200} = \begin{pmatrix} 0 & 0 \\ 0 & 0 \end{pmatrix}, \quad c_{200} = \begin{pmatrix} 0 \\ 0 \end{pmatrix}. \tag{B.30}$$

∘ $n = 1$:

∘ $s_z = -1$:

$$Q_{21-1} = \begin{pmatrix} 0 & B^{\dagger}_{211} C_{21-1} \\ 0 & C^{\dagger}_{21-1} C_{21-1} \end{pmatrix}, \quad c_{21-1} = \begin{pmatrix} 0 \\ A^{\dagger}_{200} B_{21-1} \end{pmatrix}. \tag{B.31}$$

∘ $s_z = 1$:

$$Q_{211} = \begin{pmatrix} B^{\dagger}_{211} B_{211} & 0 \\ C_{21-1} B_{211} & 0 \end{pmatrix}, \quad c_{211} = \begin{pmatrix} 0 \\ A^{\dagger}_{200} B_{211} \end{pmatrix}. \tag{B.32}$$

∘ $n = 2, s_z = 2$:

$$Q_{222} = \begin{pmatrix} D^{\dagger}_{222} D_{222} & 0 \\ 0 & D^{\dagger}_{222} D_{222} \end{pmatrix}, \quad c_{222} = \begin{pmatrix} C^{\dagger}_{21-1} D_{222} \\ B^{\dagger}_{211} D_{222} \end{pmatrix}. \tag{B.33}$$

Here, we defined Q and c for both spins, but during the flow each sector has to be treated separately, since they do not evolve with the same equations. Thus, each entry of Q defined here (it is hermitian so the upper right block can be computed for the lower left) corresponds to a block that will evolve and grow (by one row and column at each step) during the flow. It is similar for c, spin up and down must evolve separately. This initialization ends this appendix, that provided all details for an NRG implementation of the SIAM, with particle and spin conservation.

Appendix C
RGNO—Symmetry Breaking by the Few-Body Ansatz

It is shown in this appendix how the few-body ansatz for the IRLM wave function induces a spurious breaking of the PH symmetry. The wave function as written in Eq. (4.1) is recalled here:

$$|\Psi\rangle = \prod_{\alpha=-N/2}^{-M/2-1} q_\alpha^\dagger \sum_S \Psi_S |S\rangle , \tag{C.1}$$

where $|S\rangle$ are Slater deerminants concerning only correlated orbitals. The effective Hamiltonian concerning the $M+2$ correlated orbitals at any step of the iterative process, is recalled from Eq. (3.16) (in a compact form):

$$\mathcal{H}_{\text{few}} = \sum_{\substack{n,m \\ =-M/2-1}}^{M/2+1} t_m^n q_n^\dagger q_m + \sum_{\substack{n,m,p,q \\ =-M/2-1}}^{M/2+1} U_{pq}^{nm} q_n^\dagger q_m^\dagger q_p q_q , \tag{C.2}$$

where t_m^n contains every initial hopping term plus the Hartree shift induced by the uncorrelated orbitals. The interaction term is firstly investigated to verify that it respects PH symmetry in this form. The effect of the particle-hole conjugation operator P on natural orbitals (for each sector) reads $P^\dagger q_a P = q_{-a}^\dagger$. Applying this operator on the interaction term leads to (the identity $P^\dagger P = \mathbb{1}$ is inserted between each fermion operator):

M. Debertolis, *Quantum Impurity Problems in the Framework of Natural Orbitals*, Springer Theses, https://doi.org/10.1007/978-3-031-47233-6

$$
\begin{aligned}
P^{\dagger}\left(\sum_{\substack{n,m,p,q\\=-M/2-1}}^{M/2+1} U^{nm}_{pq}\, q^{\dagger}_{n} q^{\dagger}_{m} q_{p} q_{q}\right) P &= \sum_{\substack{n,m,p,q\\=-M/2-1}}^{M/2+1} U^{nm}_{pq}\, q_{-n} q_{-m} q^{\dagger}_{-p} q^{\dagger}_{-q} \\
&= \sum_{\substack{n,m,p,q\\=-M/2-1}}^{M/2+1} U^{nm}_{pq}\, q_{-n}\left(\delta_{mp} - q^{\dagger}_{-p} q_{-m}\right) q^{\dagger}_{-q} \\
&= \sum_{\substack{n,m\\=-M/2-1}}^{M/2+1} \left(U^{nm}_{mn} - U^{nm}_{nm}\right) - \sum_{\substack{n,m,p\\=-M/2-1}}^{M/2+1} \left(U^{-p-m}_{-p-n} + U^{-m-p}_{-n-p} - U^{-m-p}_{-p-n} - U^{-p-m}_{-n-p}\right) q^{\dagger}_{n} q_{m} \\
&+ \sum_{\substack{n,m,p,q\\=-M/2-1}}^{M/2+1} U^{-q-p}_{-m-n}\, q^{\dagger}_{n} q^{\dagger}_{m} q_{p} q_{q},
\end{aligned}
\tag{C.3}
$$

where indices have been changed from $-a$ to a so that fermionic operators are sorted in the proper way. If PH symmetry holds, the following equality must hold: $U^{-q-p}_{-m-n} = U^{nm}_{pq}$, that would validate the quartic term. Q is diagonalized by the orthogonal matrix D, such that $c_i = \sum_{a=-N/2}^{N/2} D_{ia} q_a$, which leads to (it is recalled that $c^{\dagger}_{-1} \equiv d$ creates a fermion in the impurity level, and the eigenvalues of Q are real such that $D^{\dagger}_{ia} = D_{ia}$):

$$
U^{-q-p}_{-m-n} = U\, D_{-1-q} D_{0-p} D_{0-m} D_{-1-n}. \tag{C.4}
$$

With the additional property $P^{\dagger} c_i P = (-1)^i c^{\dagger}_i$, it is straightforward to see that $D_{-1-a} = -D_{-1a}$ and $D_{0-a} = D_{0a}$ and hence:

$$
U^{-q-p}_{-m-n} = U\, D_{-1q} D_{0p} D_{0m} D_{-1n} = U\, D_{-1n} D_{0m} D_{0p} D_{-1q} = U^{nm}_{pq}. \tag{C.5}
$$

Therefore the quartic term obeys PH symmetry, and quadratic terms involving U must be investigated. Each term quadratic in the fermion operators that is proportional to the interaction strength in the Hamiltonian must be accounted for to consider eventual cancellations. All these terms, when transformed by the PH conjugation operator, read:

$$
\begin{aligned}
&\sum_{\substack{n,m\\=-M/2-1}}^{M/2+1} \left[-\frac{U}{2}\left(D_{-1n} D_{-1m} + D_{0n} D_{0m}\right) + \sum_{\alpha=-N/2}^{-M/2-2} \left(-U^{n\alpha}_{m\alpha} + U^{n\alpha}_{\alpha m} + U^{\alpha n}_{m\alpha} - U^{\alpha n}_{\alpha m}\right)\right] P^{\dagger} q^{\dagger}_{n} q_{m} P \\
&= \sum_{\substack{n,m\\=-M/2-1}}^{M/2+1} \left[\frac{U}{2}\left(D_{-1n} D_{-1m} + D_{0n} D_{0m}\right) + \sum_{\alpha=-N/2}^{-M/2-2} \left(-U^{n\alpha}_{m\alpha} + U^{n\alpha}_{\alpha m} + U^{\alpha n}_{m\alpha} - U^{\alpha n}_{\alpha m}\right)\right] q^{\dagger}_{n} q_{m},
\end{aligned}
\tag{C.6}
$$

up to a constant term C that is not written above. Putting Eqs. (C.3) and (C.6) together, the *interaction part* (constituted of every term proportional to U) of the Hamiltonian becomes:

$$
\begin{aligned}
P^{\dagger}\mathcal{H}_U P = \sum_{\substack{n,m\\=-M/2-1}}^{M/2+1} \Bigg[& \frac{U}{2}\left(D_{-1n}D_{-1m} + D_{0n}D_{0m}\right) \\
& + \sum_{\alpha=-N/2}^{M/2+1} \left(- U^{n\alpha}_{m\alpha} + U^{n\alpha}_{\alpha m} + U^{\alpha n}_{m\alpha} - U^{\alpha n}_{\alpha m}\right)\Bigg] q^{\dagger}_n q_m + U^{nm}_{pq} q^{\dagger}_n q^{\dagger}_m q_p q_q + C,
\end{aligned}
\tag{C.7}
$$

in such a way that it is required for PH symmetry that the following equality holds:

$$
\begin{aligned}
U\left(D_{-1n}D_{-1m} + D_{0n}D_{0m}\right) &= \sum_{a=-N/2}^{M/2+1} \left(- U^{na}_{ma} + U^{na}_{am} + U^{an}_{ma} - U^{an}_{am}\right) \\
&= \sum_{a=-N/2}^{M/2+1} U\Big(- D_{-1n}D_{0a}D_{0m}D_{-1a} + D_{-1n}D_{0a}D_{0m}D_{-1a} \\
&\quad + D_{-1n}D_{0a}D_{0m}D_{-1a} - D_{-1n}D_{0a}D_{0m}D_{-1a}\Big).
\end{aligned}
\tag{C.8}
$$

The orthonormality of the complete set of natural (and Wannier) orbitals implies the following relation: $\sum_{a=-N/2}^{N/2} P_{ia}P_{ja} = \delta_{ij}$. In Eq. (C.8), sums end at $M/2+1$, such that orthonormality relations are not fulfilled exactly. This induces an additional term in $P^{\dagger}\mathcal{H}_U P$, which is responsible for the breaking of PH symmetry.

Nevertheless, in regimes where Kondo correlations are small, natural orbitals are expected to be close to Wannier orbitals as seen in Fig. 3.8, and most of the weight of sites -1 and 0 is carried by correlated orbitals such that $\sum_{a=-N/2}^{N/2} P_{ia}P_{ja} \simeq \delta_{ij}$. The symmetry breaking is then quantitatively small, and the occupancy of the dot stays exponentially close to $1/2$. On the other hand, when correlations increase in the system, the impurity is diluted among more and more natural orbitals (as are other sites of the chain) and the symmetry breaking gets bigger and bigger. The impurity is therefore more polarized ($n_d - 1/2$ can become of order 10^{-2}), which is similar to an infrared cutoff $> T_{\mathrm{K}}$ that prevents correlations to fully develop. Forcing Eq. (C.8) to be exact has been tested in actual simulations and lead to numerical instabilities preventing any convergence of the algorithm. Indeed, it changes in some sense the parameters of the true Hamiltonian and therefore the targeted ground state changes at every iteration, while the algorithm is really sensitive to such inconsistencies.

Appendix D
RGNO—Details on the Exact Diagonalization

Details concerning the implementation of the exact diagonalization (ED) of the effective Hamiltonian (dealing with $M + 2$ correlated orbitals) performed at each RGNO iteration are given in this appendix. It is not claimed that the proposed method is optimal, but has proven to be efficient enough for the purpose of the studies performed with the RGNO. The code has been fully implemented in Python, with help of the external libraries Numpy and Scipy, providing optimized C-based implementations of various linear algebra operations. Note that the efficient broadcasting methods of arrays implemented in the Numpy library overpass time consuming operations with lists inherent to basic Python codes.

D.1 Basis Representation and Fock Space

The exact diagonalization of an Hamiltonian describing interacting particles needs to be performed in the Fock space, whose basis contains all possible states. In the present case, the effective Hamiltonian projected in the correlated sector containing $M + 2$ orbitals can be simply written:

$$\mathcal{H}_{\text{few}} = \sum_{n,m=0}^{M+1} t_m^n q_n^\dagger q_m + \sum_{n,m,p,q=0}^{M+1} U_{pq}^{nm} q_n^\dagger q_m^\dagger q_p q_q + C, \tag{D.1}$$

where hoppings and interactions are *all-to-all*, and C is a constant shift. This spinless Hamiltonian contains 2^{M+2} states, corresponding to all combinations from zero to $M + 2$ particles in $M + 2$ orbitals, with at most 1 particle per orbital (Pauli principle). The above Hamiltonian enjoys the conservation of the number of particles, such that it can be diagonalized by blocks of fixed number of particles. Since the solution is expected to be close to the PH symmetric one, the ground state is expected to be found

M. Debertolis, *Quantum Impurity Problems in the Framework of Natural Orbitals*, Springer Theses, https://doi.org/10.1007/978-3-031-47233-6

in the half-filled sector. Without the need of the full spectrum, only this sector will be diagonalized, and its ground state can be accurately found by a Lanczos procedure (routines for the Lanczos are taken from the library Scipy, suited for sparse matrices). The Lanczos algorithm is very profitable when only the first few (or last) eigenvalues and eigenvectors are desired, which corresponds exactly to the need of the RGNO: at each iteration, only the lowest energy state is required to project the correlators defining the one-body density matrix.

Fock states can be defined differently according to the chosen representation. Here, the sector of fixed number of particles needs to be clearly defined, such that the occupancy representation of states is chosen. A given state is hence written $|\Psi\rangle = |n_0, \ldots, n_{M+1}\rangle$, where $n_i \in \{0, 1\}$ is the occupancy of the ith correlated orbital. In the text, correlated orbitals were indexed from $-M/2 - 1$ to $M/2 + 1$, but indices will be shifted here by $M/2 + 1$ to range from 0 to $M + 1$ for convenience. The half-filled sector is thus defined (M chosen even) by each state in which $\sum_{i=0}^{M+1} n_i = (M + 2)/2$. In the following, $M_2 = M + 2$ is written to lighten the notation.

Numerically, a fermionic state can be efficiently defined through binary numbers: for instance, state $|\Psi\rangle = |010110\rangle$ (for $M_2 = 6$) can be stored by the binary number $\overline{010110}^2 = \overline{26}^{10}$. Each number between 0 and 2^{M_2} has its own binary notation, and the number of particles in each state is the sum of ones in the corresponding binary number. Here, it is (arbitrarily) chosen to count the powers of 2 from the left to the right, which will be the convention in the following: $\overline{100000}^2 = 1$, $\overline{000001}^2 = 32$. States range from $\overline{000000}^2 = 0$ to $\overline{111111}^2 = 2^{M_2} - 1$ which correctly generates 2^{M_2} states. In the half-filled sector, there are $\mathcal{N}_{\text{h-f}} = \binom{M_2}{M_2/2} = \frac{M_2!}{(M_2/2)!^2}$ states only, which corresponds to a reduction of the number of states by a factor ~ 3 for $M_2 = 6$, and ~ 5 for $M_2 = 12$.

D.2 Sparse Representation: Masks

Once states in the half-filled sector are defined, the Hamiltonian must be constructed in the given basis. The form of Eq. (D.1) is very general, and no additional symmetries can be implemented. However, a lot of states do not couple through the Hamiltonian, i.e. $\langle\Psi_a|\mathcal{H}_{\text{few}}|\Psi_b\rangle = 0$, which renders the matrix in the Fock space relatively sparse. The Hamiltonian can change the occupancies of 0, 2 or 4 orbitals in a given state (without changing its number of particles). For instance, the state $|010011\rangle$ can not be transformed to the state $|101100\rangle$ by application of the Hamiltonian. When two states do not *couple*, the corresponding entry in the Hamiltonian matrix is zero, and bears some value otherwise. Actually, there are three ways for two states to couple: they can be identical (no orbital changed its occupancy), a single 1 in a state can exchange to an empty orbital (a zero in the binary notation) (two orbitals changed their occupancies), or two 1s can exchange to two 0s (four orbitals changed their occupancies). The knowledge of the coupling matrices between each state allows to compute only non-zero elements of the Hamiltonian, which is a gain of computational time and of storage. Each entry is stored as (i, j, H_{ij}), and we only need to define every state j coupling to each state i (the Hamiltonian is hermitian, such that we define only the upper (or lower) part of the matrix, $j < i$), and then evaluate the corresponding Hamiltonian element (see the following Sect. D.3). Hence, there are three masks to define for each coupling possibility between two states presented above.

1. The first one is straightforward as it corresponds to diagonal elements of the Hamiltonian $i = j$.

2. For the second one and the third one, two matrices F_{ia} and E_{ia} are introduced: F_{ia} contains the position in state i of each occupied (or empty for E) orbital. Thus, $i \in [0, \mathcal{N}_{\text{h-f}}[$ and $a \in [0, M_2[$, and at half-filling both matrices are of dimension $(\mathcal{N}_{\text{h-f}}, M_2/2)$. Then, the binary representation of the state is used to define all states to which it couples: if the state i corresponds to $N_i = \overline{010110}^2 = 26$, then it couples at the state represented by $N_j = \overline{010011}^2 = 50$ through operators $q_5^\dagger q_3$ of the Hamiltonian. In this case, $F_i = (1, 3, 4)$ and $E_i = (0, 2, 5)$. Hence, the operation used to find the state 50 is $N_j = N_i + 2^{E_{i,a}} - 2^{F_{i,b}}$, where $(a, b) = (5, 3)$ in this example. In this way, each pair (a, b) generates a new state to which state i couples, if $a \neq b$. If $a = b$, the state is the same ($N_i = N_j \rightarrow i = j$) and is already defined through the first mask. However, $F_i \cap E_i = 0$, such that $a \neq b$ is always true. To avoid double counting of states (interchanging a and b leads to the same state), the condition $a < b$ is imposed.

3. For the third way of coupling, the same method is used yet involves more indices: $N_j = N_i + 2^{E_{i,a}} + 2^{E_{i,b}} - 2^{F_{i,c}} - 2^{F_{i,d}}$, with $a \neq b$ and $c \neq d$ to respect the Pauli principle. As before, $a < b$ and $c < d$ are imposed to avoid multiple counting of the same states.

Since states are defined in the half-filled (or any) sector, N_j does not correspond to the jth entry of the Hamiltonian in this sector. It is thus necessary to build a list K of size $\mathcal{N}_{\text{h-f}}$ with entries $K_{N_j} = j$, thanks to which the masks are properly defined. With these masks, the number of matrix elements is reduced from $\mathcal{N}_{\text{h-f}}^2$ to:

$$\mathcal{N}_{\text{sparse}} = \mathcal{N}_{\text{h-f}} \left(1 + \left(\frac{M_2}{2} \right)^2 + \frac{\frac{M_2}{2}!}{2!(\frac{M_2}{2} - 2)!} \right), \tag{D.2}$$

which allows to consider only $\sim 40\%$ of the states for $M_2 = 8$, 10% for $M_2 = 12$ and even less for larger M.

D.3 Construction of the Hamiltonian

The masks being created, each non-zero entry of the Hamiltonian has to be filled by its value. For each set of coordinate (i, j) of each mask, different terms of the Hamiltonian are involved, and each of the three different cases are treated separately. Since the system is constituted of fermionic particles, the fermionic sign arising from anti-commutation relations has to be treated with care for each matrix element. In order to compute the fermionic sign, different options are possible: for $\langle\Psi_j|q_a^\dagger q_b|\Psi_i\rangle$ (with $n_a = 0$ and $n_b = 1$ in $|\Psi\rangle$), the correct sign of the term can be the sign of the permutation between a and b in the sequence defined by $|\Psi_i\rangle$. It can also be defined by the power of (-1) given by the number of fermions *passed* by q_b and $q_a^\dagger$. The first method requires the sign of each permutation, which is the determinant of the corresponding permutation matrix. The second method is on the contrary easily implemented, and is prefered in the actual code.

1. Diagonal terms concern the constant part of the Hamiltonian, plus every term that does not change the state after application. Hence, hopping elements contribute by $\sum_{n_a=1} t_a^a$ through operators $q_a^\dagger q_a$, if $a \in F_i$ (i.e. orbital a is filled in the state). The interaction term also participates by an amount $\sum_{n_a=n_b=1}(U_{ba}^{ab} - U_{ab}^{ab})$, with $a \in E_i$, $b \in E_i$ and $a \neq b$.

2. Terms defined through the second mask are every hopping terms $s_{2-ab} t_b^a$ ($a \neq b$, and $s_2 = \pm 1$ is the fermionic sign) such that $a \in E_i$ and $b \in F_i$, and interaction terms with two equal indices that are summed $s_2 \sum_{n_b=1}(U_{bc}^{ab} + U_{cb}^{ba} - U_{cb}^{ab} - U_{bc}^{ba})$, with $a \in E_i$, $b \in F_i$, $c \in F_i$, $b \neq c$.

3. Remaining matrix elements defined by the third mask are only coming from the interaction term involving four different orbitals, $s_{4-abcd}(U_{cd}^{ab} + U_{dc}^{ba} - U_{cd}^{ba} - U_{dc}^{ab})$, with $s_4 = \pm 1$.

Diagonal terms have no fermionic sign because operators $q_a^\dagger q_a$ come always in pair, such that if q_a yield to a minus sign, it will be canceled by the same minus sign brought by $q_a^\dagger$. The two other fermionic signs are given by:

$$
\begin{aligned}
s_{2-ab} &= (-1)^{\sum_{i<a} n_i + \sum_{i<b} n_i} \times \text{sign}(b-a) \\
s_{4-abcd} &= (-1)^{\sum_{i<a} n_i + \sum_{i<b} n_i + \sum_{i<c} n_i + \sum_{i<d} n_i} \times \text{sign}(d-c) \times \text{sign}(a-b),
\end{aligned}
\tag{D.3}
$$

where sign functions restore the correct sign if $b < a$, since $\sum_{i<a} n_i$ counts the b orbital while it was emptied by q_b before.

D.4 One-Body Density Matrix

At each iteration, the one-body density matrix is constructed in the ground state of the current enlarged Hamiltonian. In the uncorrelated sector, the one-body density matrix Q is trivially 1 or 0, if the orbital is in the occupied or unoccupied sector respectively. Only the block concerning the $M + 2$ correlated orbitals is relevant in order to obtain the new set of natural orbitals. From the exact diagonalization of the sparse Hamiltonian by Lanczos, the transformation matrix going from the natural orbital basis to the ground state is used to project each correlator $\langle q_n^\dagger q_m \rangle$. Each of these correlators is defined in the Fock space in the same manner as the Hamiltonian, with only one element $q_n^\dagger q_m$ of unique amplitude 1.

Any static observable at $T = 0$ involving no more than 4 fermionic operators can be defined similarly: it is the case for the occupancy of the dot (that is actually the first element of the Q-matrix) or the screening cloud correlator C_{ij}.

D.5 Variance of the Hamiltonian

The variance is computed to determine quantitatively the accuracy of a given state. Within the few-body wave function, a part of this observable is projected in the Fock space, as the one-body density matrix previously. However, the variance (that is defined in Eq. (4.2)) contains higher order terms that require an extension of the construction of matrices in Fock space. To take advantage of the shape of the interaction in real space, the Hamiltonian is squared in the Wannier representation:

$$(\mathcal{H}_{\text{IRLM}})^2 = U^2 \left(d^\dagger c_0^\dagger c_0 d\right)^2 + \sum_{ij \geq -1} t_j^i c_i^\dagger c_j (d^\dagger c_0^\dagger c_0 d) + H_{\text{quart}} + H_{\text{quadra}}, \quad \text{(D.4)}$$

where t_j^i concerns every quadratic term of $\mathcal{H}_{\text{IRLM}}$, and H_{quart} (H_{quadra}) contains every quartic (quadratic) term arising when taking the power of the Hamiltonian, which do not require any extension of the algorithm.

Fortunately (when it is not a matter of fermionic sign) the particles are fermions and the property $(d^\dagger d)^2 = d^\dagger d$ (similarly $(c_0^\dagger c_0)^2 = c_0^\dagger c_0$) reduces the first term to a quadratic term, so that only terms dealing with 6 fermionic operators are to implement. A new mask is built for the operators linking two states that differ by 3 fermions (note that this is not necessary for $M_2 \leq 4$, but actual simulations consider at least $M_2 \geq 6$), then the corresponding terms of $\mathcal{H}$ are determined.

Mask

As before, the states that couple to each other are found through the binary representation of states $N_j = N_i + 2^{E_{i,a}} + 2^{E_{i,b}} + 2^{E_{i,c}} - 2^{F_{i,d}} - 2^{F_{i,e}} - 2^{F_{i,f}}$, with $c < a$, $c < b$, $b < a$, $d < f$, $d < e$ and $e < f$. The number of states coupling to each state through this term is then:

$$\mathcal{N}_3 = \binom{\frac{M_2}{2}}{3} = \frac{\frac{M_2}{2}!}{3!(M_2 - 3)!}, \quad \text{(D.5)}$$

and for $M = 10$ this number equals $\mathcal{N}_3 = 400$, that is only half of $\mathcal{N}_{\mathrm{h-f}} = 924$ (here the hermitian character of the operator is not accounted) which makes the calculation of the variance expensive for $M \geq 12$. However, the exponential decay of the variance allows us to compute it for smaller M, and to trust results for larger M if they were already converged at a reasonable precision.

Matrix Elements

The matrix elements will spread into the 4 masks depending on the indices of the tensors. The 6-legs tensor is build from: $V_{abcdef} = \sum_{i,j=-1}^{N-1} t^i_j D_{ia} D_{jf} U_{bcde}\ q^\dagger_a q^\dagger_b q^\dagger_c q_d\ q_e q_f$. Then, as before, one needs to take every term depending on the number of equal indices. For no equal indices, there is one term linking i to j, 9 for two equal indices (j belongs to the third mask), 18 for four equal indices (second mask), and 6 for all indices equal by pairs (first mask). Each fermionic index between state i and j, and the relative signs between every term must also be computed carefully. The calculation of the variance opens the possibility to compute every observable that contains 6 fermionic operators, but usually physical observables of interest have no more than 4 operators.

Appendix E
RMT—Calculating Distribution Functions in Simpler Models

The calculations performed in the last section of Chap. 6 are reproduced here for simpler random matrix models, namely the free problem and the one with a rank one perturbation on this free problem.

E.1 Random Model Without Impurity

This is the simplest model that can be built, as the corresponding Hamiltonian has all its entries randomly distributed:

$$P\left(H_{ij}\right) = \frac{1}{\sqrt{2\pi}\sigma} \exp\left(-\frac{1}{2\sigma^2} H_{ij}^2\right), \tag{E.1}$$

and the joint distribution of its eigenvalues and eigenvectors are given by the Porter-Thomas distribution:

$$P_{\text{P-T}}(\{e_\alpha\}, \{r_\alpha\}) = \prod_{1 \le \alpha < \beta}^{N} |e_\alpha - e_\beta| \prod_{1=\alpha}^{N} \frac{1}{\sqrt{r_\alpha}} \delta\left(\sum_{\beta=1}^{N} r_\beta - 1\right) e^{-\frac{1}{2\sigma^2} \sum\limits_{\beta=0}^{N} e_\beta^2}. \tag{E.2}$$

The density of energies in this model are known, and are given by the Wigner's semi-circle law:

$$u(e) = \frac{\sqrt{2\sigma^2 N - e^2}}{2\pi\sigma^2}, \quad \text{with} \quad \int\limits_{-\sqrt{2N}\sigma}^{\sqrt{2N}\sigma} \mathrm{d}e\, u(e) = N. \tag{E.3}$$

The quantity that we calculate here will be the occupancy of a given *site* n_i (note that this quantity is the same for any site i in this problem), even if no spatial scale

M. Debertolis, *Quantum Impurity Problems in the Framework of Natural Orbitals*, Springer Theses, https://doi.org/10.1007/978-3-031-47233-6

is present in this problem. This quantity is studied to make the parallel with the occupancy of the impurity in the model presented in the main text. The occupancy of a given *site i* is defined as follows:

$$n_i = \sum_{E_\alpha < 0} r_\alpha . \tag{E.4}$$

Following the simplifications presented in Sect. 6.5.1, and by inserting the semi-circle law for the density of levels, one can define the distribution of this occupancy in a convenient form:

$$P(n_i) = \frac{1}{\mathcal{Z}} \int_{-\infty i}^{+\infty i} \mathrm{d}\lambda \int_{-\infty i}^{+\infty i} \mathrm{d}\mu \, e^{-NF\left(\lambda,\mu,n_i\right)}, \tag{E.5}$$

where all constant terms (that does neither depend on the Lagrange multipliers nor on n_i) are absorbed in the norm $\mathcal{Z}$, and the functional in the exponential reads:

$$F\left(\lambda, \mu, n_i\right) = \frac{1}{2N} \int_0^{\sqrt{2N}\sigma} \mathrm{d}e \, u(e) \ln(\lambda) + \frac{1}{2N} \int_{-\sqrt{2N}\sigma}^{0} \mathrm{d}e \, u(e) \ln(\lambda + \mu) - \frac{1}{N} \left(\mu n_i + \lambda\right). \tag{E.6}$$

Every term in this functional is expected to be of order 1, since the density of states $u(e)$ integrates to N. Hence, the factor N can be explicitely factorized in Eq. (E.5), and the steepest descent method, or saddle-point approximation, can be used to compute the integrals over the multipliers. This method replaces the value of the integral by the value of the function at its maximum, that is dominant due to the power N in the exponential. The saddle-point equations are easily solved for this model, as the integrals on the semi-circle are trivially evaluated:

$$\begin{cases} \partial_\lambda F = 0 \Rightarrow \int_0^{\sqrt{2N}\sigma} \mathrm{d}e \, \frac{u(e)}{2\lambda} = 1 - n_i \, , \\ \partial_\mu F = 0 \Rightarrow \int_{-\sqrt{2N}\sigma}^{0} \mathrm{d}e \, \frac{u(e)}{2(\lambda+\mu)} = n_i \, , \end{cases} \Rightarrow \begin{cases} \lambda = \frac{N}{4} \frac{1}{(1-n_i)}, \\ \mu = \frac{N}{4} \frac{1-2n_i}{n_i(1-n_i)}. \end{cases} \tag{E.7}$$

The integrals over λ and μ in Eq. (E.5) are replaced by the functional evaluated at the saddle-points, and the integration over the semi-circle leads to a simple form for the distribution of occupancies:

$$\boxed{P(n_i) = \frac{1}{\mathcal{Z}} \left(n_i (1 - n_i)\right)^{N/4}}. \tag{E.8}$$

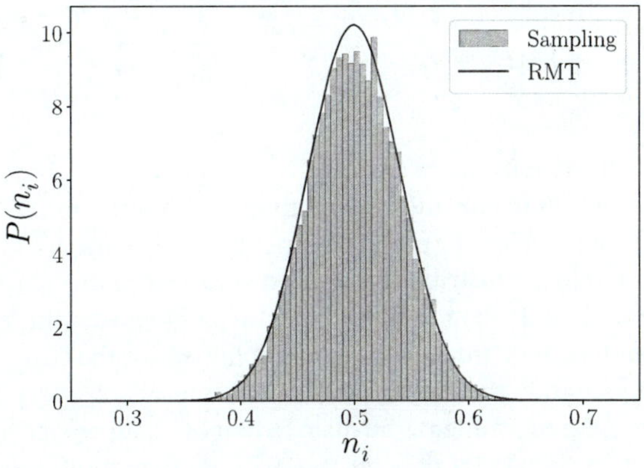

Fig. E.1 Analytical (RMT) and numerical results (sampling) for the distribution function of the occupancy of a given *site* of the free random matrix problem. These results concern matrices of dimension $N = 300$, and the dispersion of the random variables is $\sigma = \sqrt{2}$. The sampling is performed over 10^4 realizations of the random matrix

Since the dependence on n_i is determined, the norm can be computed through the normalization constraint of $P(n_i)$ between 0 and 1. This distribution is plotted in Fig. E.1 together with the distribution computed numerically, and both functions fit perfectly.

In the diluted regime of the impurity problem presented in the Chap. 6, the distribution of eigenvalues was almost the one of the free problem, and we could expect that the distribution of n_d could be similar to the one presented here. However, the coupling V/σ rules the width of this distribution in the impurity problem, such that an effective description of the diluted regime could use the result of Eq. (E.8) but with a renormalized $\sigma \propto V$, that would change the power of $N/4$ and adjust the width on the distribution in a simple manner.

E.2 Random Model with Onsite Potential

The second model that is studied here is the case of a rank-one perturbation on the free problem. This model was studied already in Ref. [1, 2] to explain deviations from the Porter-Thomas distribution observed in experimental results. The corresponding problem is the following:

$$H_{ij} = G_{ij} + Z\delta_{i1}\delta_{j1}, \tag{E.9}$$

where Z is chosen to be proportional to $\sigma\sqrt{N}$, such that the parameter Z is always relevant. In the work of Bogomolny [2], the joint distribution of eigenvalues and eigenvectors of H was found:

$$P\left(\{E_\alpha\},\{z_\alpha\}\right)=\prod_{1\le\gamma<\delta}|E_\gamma-E_\delta|e^{-\frac{1}{2\sigma^2}\sum_{\gamma=1}^{N}(E_\gamma^2-2ZE_\gamma z_\gamma)}\prod_{\gamma\ge1}\frac{1}{(z_\gamma)^{\frac{1}{2}}}\delta\left(1-\sum_{\gamma=1}^{N}z_\gamma\right), \tag{E.10}$$

in such a way that the eigenvectors and eigenvalues are no more independent because of the term $E_\alpha z_\alpha$, similar to our study concerning the impurity model. This expression is valid for the *site* onto which the perturbation is applied, namely $i=1$. In this model, we will be interested in the distribution of the occupancy of this particular site. When Z becomes greater than the bandwidth $\sqrt{2N}\sigma$, a single level with a weight z_{out} gets out of the band and must be treated apart, as explained for the two outliers presented in Chap. 6. First, we study the case of $Z<\sqrt{2N}\sigma$, in which every z_α is of the same order $1/N$. As in the free problem, the distribution of occupancies is expressed as an exponential of a functional that depends on the Lagrange multiplier and the wanted occupancy number n_1. This functional reads:

$$F\left(\lambda,\mu,n_1\right)=\frac{1}{2N}\int_0^{\sqrt{2N}\sigma}\mathrm{d}E\,u(E)\ln\left(|\lambda-\frac{Z}{\sigma^2}E|\right)+\frac{1}{2N}\int_{-\sqrt{2N}\sigma}^{0}\mathrm{d}E\,u(E)\ln\left(|\lambda+\mu-\frac{Z}{\sigma^2}E|\right)$$
$$-\frac{1}{N}\left(\mu n_1+\lambda\right). \tag{E.11}$$

This expression is not as simple as before as the logarithms depends on the energy, which complicates analytical calculations. Note that as before, the density of states at first order stays the semi-circle. The saddle-point equations to be solved for the two multipliers are given by the transcendental equations:

$$\begin{cases}\frac{1}{\pi Z}\left(1+\frac{\pi\lambda}{2Z}-2\,\mathrm{aCot}\left[\frac{\lambda/V-1}{\sqrt{(\lambda/Z)^2-1}}\right]\sqrt{\left(\frac{\lambda}{Z}\right)^2-1}\right)=1-n_1\\ \frac{1}{\pi Z}\left(-1+\frac{\pi(\lambda+\mu)}{2Z}-2\,\mathrm{Re}\left[\mathrm{aCosh}\left[\frac{Z}{\lambda+\mu}\right]\sqrt{\left(\frac{Z}{\lambda+\mu}\right)^2-1}\right)\right]=n_1,\end{cases} \tag{E.12}$$

which can not be solved analytically. The complications comes from the fact that the Heaviside function cuts the integration over the energies in two separates parts, which brings after integration the trigonometric functions. If the integration can be performed on the whole range of energies $[-\sqrt{2N}\sigma,\sqrt{2N}\sigma]$, simpler results can be found for the integrals using complex analysis.

Therefore, the solutions to these two equations are found numerically, and the corresponding values of λ and μ are reinjected in the functional F. Then, the last integrals over E in F are also computed numerically, and we obtain the wanted distribution $P(n_1)$. This distribution is plotted for two different values of Z in Fig. E.2, and reproduce quantitatively the numerical simulation. As expected, the site is depleted for Z positive, and the symmetric behavior with respect to $n_1=0.5$ is found for Z negative, since in this case the site is more favorable to be occupied.

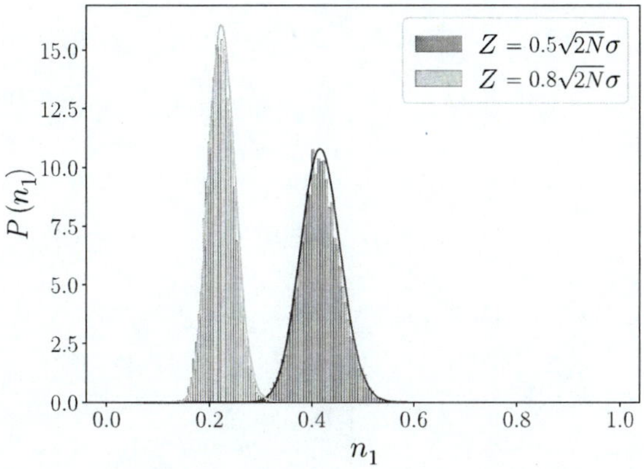

Fig. E.2 Analytical (lines) and numerical (bars) solutions for the distribution of the occupancy of the site containing the perturbation, for two different perturbation strengths Z. As before, $N = 300$, $\sigma = \sqrt{2}$ and 10^4 realizations of the random matrix are simulated for the sampling. As Z increases, the average occupancy is lower, as is the width of the distribution

When Z becomes stronger than the bandwidth, the calculations are similar, but a state must be taken out of the density of states $u(E)$. The integrals over E_{out} and z_{out} must be done in addition to the one over λ and μ, and they are also treated through saddle-point equations. The solutions for these new saddle-point equations are given in Ref. [2], such that as before, only the Lagrange multipliers are left to find. The functional in this case would read:

$$F\left(\lambda,\mu,n_1\right) = \frac{1}{2N}\int_0^{\sqrt{2N}\sigma} \mathrm{d}E\, u(E)\ln\left(|\lambda - \frac{Z}{\sigma^2}E|\right) + \frac{1}{2N}\int_{-\sqrt{2N}\sigma}^{0} \mathrm{d}E\, u(E)\ln\left(|\lambda+\mu-\frac{Z}{\sigma^2}E|\right)$$
$$+\frac{1}{N}\left(\ln\left(|E_{\text{out}}|\right) - \frac{E_{\text{out}}^2}{2\sigma^2} - \mu\left(n_1 - \Theta(-E_{\text{out}})z_{\text{out}}\right) + \lambda(1-z_{\text{out}})\right), \tag{E.13}$$

The numerical solution of this problem is similar to the one of the previous situation, and is not shown here as it brings no additional information concerning the analytical calculation of distribution functions in this type of problems.

References

1. Aleiner IL, Matveev KA (1998) Shifts of random energy levels by a local Perturbation. en. Phys Rev Lett 80:814–816. ISSN: 0031-9007, 1079-7114
2. Bogomolny E (2017) Modification of the Porter-Thomas distribution by rank-one interaction. en. Phys Rev Lett 118:022501. ISSN: 0031-9007, 1079-7114